I0030224

Mines of
Tuscarora, Cortez, Seven Troughs and other Northern Nevada Districts

by W. H. Emmons

Notes on Some Mining Districts in
Humboldt [Pershing] County
by F. L. Ransome

This is a photographic reproduction of the 1910 U.S. Geological Survey Bulletin 408 entitled "A Reconnaissance of Some Mining Camps in Elko, Lander and Eureka Counties, Nevada," together with the 1909 U.S. Geological Survey Bulletin 414 entitled "Notes on Some Mining Districts in Humboldt County, Nevada."

Published by Miningbooks.com
For
Stanley Paher
Nevada Publications

CONTENTS.

ILLUSTRATIONS.

8

A RECONNAISSANCE OF SOME MINING CAMPS IN ELKO, LANDER, AND EUREKA COUNTIES, NEVADA.

By William H. Emmons.

INTRODUCTION.

SCOPE OF REPORT.

The following report gives the results of a reconnaissance of the mining districts in northeastern Nevada which are included between the one hundred and sixteenth and one hundred and seventeenth meridians and the fortieth and forty-second parallels. The eastern boundary of the area crosses the Bullion and Lone Mountain mining districts, and is about 12 miles west of Elko and 6 miles east of Carlin; the northern boundary is the Idaho line; and the western boundary passes about 12 miles west of the Gold Circle district and about 3 miles west of Battle Mountain. The general outline of the area and its position with respect to neighboring places are shown on the index map (fig. 1). As thus defined it includes about 7,000 square miles, embracing the western part of Elko County and the northern parts of Lander and Eureka counties. The report includes brief descriptive notes on the geology and ore deposits of the following camps: Gold Circle (Midas), Tuscarora, Lime Mountain, Edgemont, Aura, Columbia, Mountain City, Van Duzer Creek, Cornucopia, Good Hope, Burner, Falcon, Lone Mountain, Palisade (Safford district), Bullion (Railroad district), Mineral Hill, Alpha, Lynn, Cortez, Mill Canyon, Dean, Lewis, Maysville, Pittsburg, Hilltop, Tenabo, Lander, Mud Springs, and Grey Eagle.

FIELD WORK AND ACKNOWLEDGMENTS.

The work in the field was done between July 15 and October 20, 1908. During this time all the important mining districts were visited, but no geologic mapping was undertaken except that incidental to examination of the ore deposits. The reports of the United States Geological Exploration of the Fortieth Parallel, which covered about two-thirds of the area described herein, have been of great service as an aid in the study of the ore deposits. Plate V is a generalized reproduction, with some minor changes, of a part of Plate IV from the atlas of the Fortieth Parallel Survey.

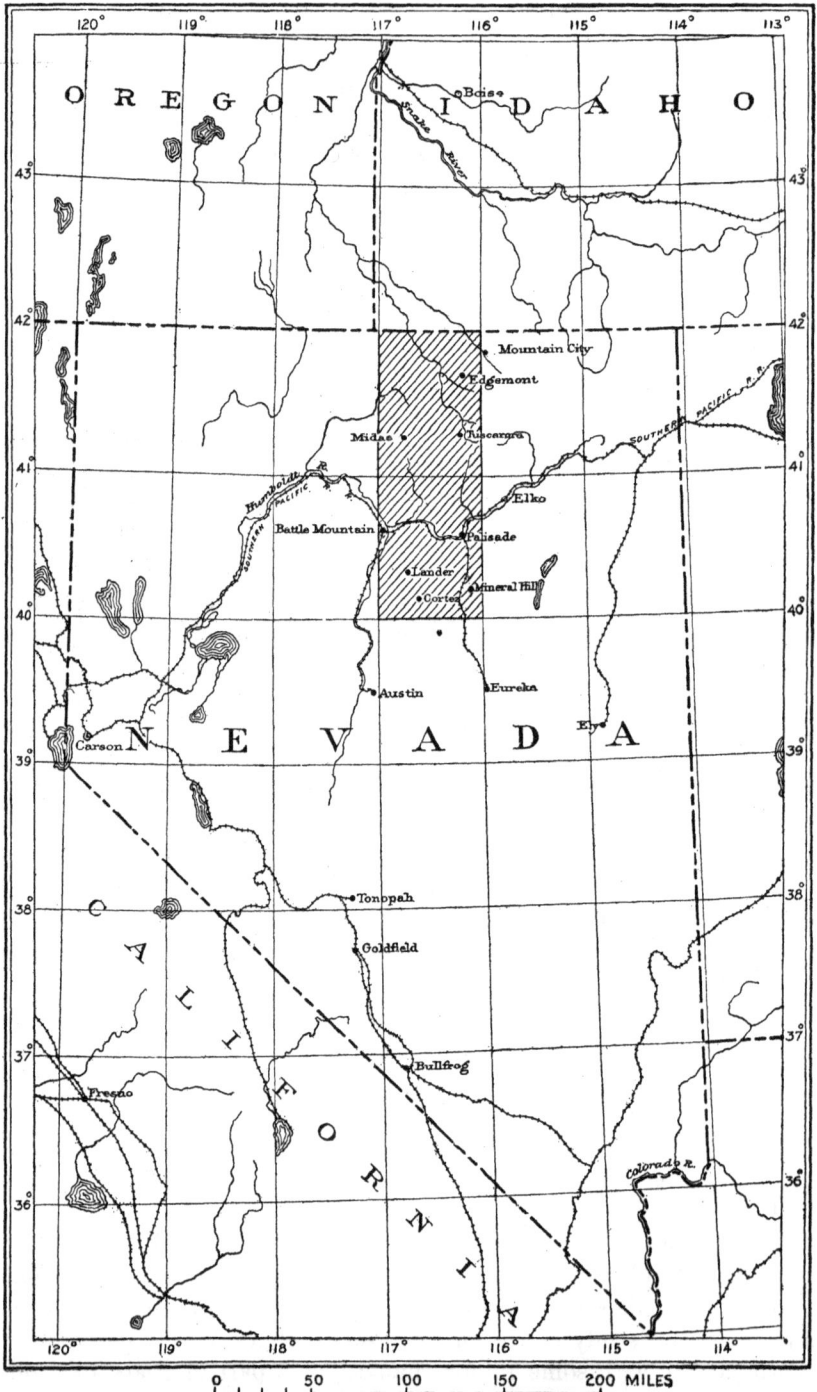

FIGURE 1.—Index map showing location of area described.

A great many of the mines were inaccessible when the camps were visited. Some of them had not been worked since the eighties, when silver mining was at high tide. The notes on such mines are of the most general character and treat mainly the surface geology and the general relations of the ore deposits.

The writer is indebted to Mr. Waldemar Lindgren, in charge of the section of metalliferous ore deposits of the United States Geological Survey, for criticism and advice, and to the mine owners, operators, and prospectors of the area, who without exception have offered every facility to aid the work.

BIBLIOGRAPHY.

The following publications include the most important literature relating to the area described and adjacent territory in Nevada. Many notes of great historical interest may be found in the early reports to the United States Government by R. W. Raymond and J. Ross Brown, and in later reports of the Bureau of the Mint. A pamphlet treating of the mineral wealth of Elko County, published in 1907 and distributed gratis by the Chamber of Commerce of Elko, Nev., contains interesting details regarding some of the mines.

BANCROFT, HUBERT HOWE. History of Nevada, p. 322. A reliable and entertaining account of the history of Nevada from the earliest settlements to 1888. Includes a narrative of the development of the Comstock lode and other important districts, and some valuable data concerning the early history of Elko, Lander, and Eureka counties.

CURTIS, J. S. Silver-lead deposits of Eureka, Nev. Mon. U. S. Geol. Survey, vol. 7, 1884. A study of the geology and ore deposits of the Eureka district, with detailed descriptions of the mines.

GILBERT, G. K. Lake Bonneville. Mon. U. S. Geol. Survey, vol. 1, 1890; also Rept. Geog. and Geol. Surveys W. 100th Mer., vol. 3, 1875. Contain some valuable generalizations respecting geology of Nevada. The area described in the present reconnaissance was not covered by Lake Bonneville.

HAGUE, ARNOLD. Geology of the Eureka district, Nevada, with an atlas. Mon. U. S. Geol. Survey, vol. 20, 1892. Describes the geology and ore deposits of the Eureka district and gives the best section of Paleozoic sedimentary rocks available for northeast Nevada.

KING, CLARENCE. Geological exploration of the fortieth parallel. This great work, including six volumes and an atlas, is the result of the joint labors of Clarence King, S. F. Emmons, Arnold Hague, James Hague, and others. The area described is a belt nearly 2° wide along the Southern Pacific and Union Pacific railroads from a point east of Cheyenne, Wyo., to a point west of Pyramid Lake, near the east front of the Sierra Nevada. It includes the central and southern portions of the area covered in the present reconnaissance, and the reports are the source of many of the geologic data given herein. References to special parts of this work are made in appropriate places.

LINDGREN, WALDEMAR. The geological features of the gold production in North America. Trans. Am. Inst. Min. Eng., vol. 33, 1902, p. 790. Discusses the precious-metal production of Nevada and suggests that Tuscarora affords a connecting link between the Tertiary veins of Nevada and those of Idaho.

LOUDERBACK, GEORGE DAVIS. Basin range structure of the Humboldt region. Bull. Geol. Soc. America, vol. 15, 1904, pp. 289–346. Concludes that the present topography of certain of the basin ranges is due to faulting, and modified to only a slight extent by erosion.

McCLELLAN, E. C. Map of Elko County, Nev. A reliable hachure map of Elko County, a part of which is reproduced herein. Published by the author, Elko, Nev.

RUSSELL, I. C. Geological history of Lake Lahontan, a Quaternary lake of northwestern Nevada. Mon. U. S. Geol. Survey, vol. 11, 1885. Contains some generalizations respecting the geology of Nevada.

SPURR, J. E. Origin and structure of the basin ranges. Bull. Geol. Soc. America, vol. 12, 1901, pp. 217–270. Includes a comprehensive review of the literature relating to the structure of the basin ranges and concludes that their present topographic expression is due mainly to erosion.

SPURR, J. E. The succession and relation of lavas of the Great Basin region. Jour. Geology, vol. 8, 1900, pp. 621–646. A summary of the age relations of the Tertiary rocks of Nevada, with some generalizations bearing on the problems of magmatic differentiation.

SPURR, J. E. Descriptive geology of Nevada south of the fortieth parallel and adjoining portions of California. Bull. U. S. Geol. Survey No. 208, 1903. An account of a reconnaissance of the area indicated by the title, which includes a review of previous work done in the area, together with a geologic map. The area borders on the south that described in the present paper.

TOPOGRAPHY, CLIMATE, AND VEGETATION.

The area here described (see Pls. I, II, III, and IV; fig. 2) is a part of the Great Basin and has the characteristic basin topography. The country is a low plateau traversed by long, narrow mountain ranges, which trend northward and are approximately parallel. Between the mountains are wide valleys parallel to the ranges. The lowest parts of the plateau are a little less than 5,000 feet above sea level and the summits of the ranges are in the main between 8,000 and 10,000 feet. The drainage of the country has a gridiron arrangement, the chief stream being Humboldt River, which flows westward across the mountain ranges to the Carson Sink, receiving tributaries at right angles from the valleys between the north-south ranges on both sides. The north third of the area is drained by Owyhee River, a tributary of Snake River, which empties into the Columbia.

The summers are long and hot, the winters are rather severe for the latitude, and the air is dry. In the summer there are frequent showers of short duration and occasional cloud-bursts. In the winter the precipitation is mainly snow, which in the highest mountains accumulates in sufficient quantity to last well into the middle of the summer. The lower country is covered by sagebrush, which grows to great size, especially in the lower parts of the valleys. High on the mountain slopes there are several varieties of pine trees, junipers, and south of Humboldt River some mountain mahogany. Farming is carried on and wild grasses are cut along the Humboldt and its tributaries, and wherever there is sufficient water for irrigation

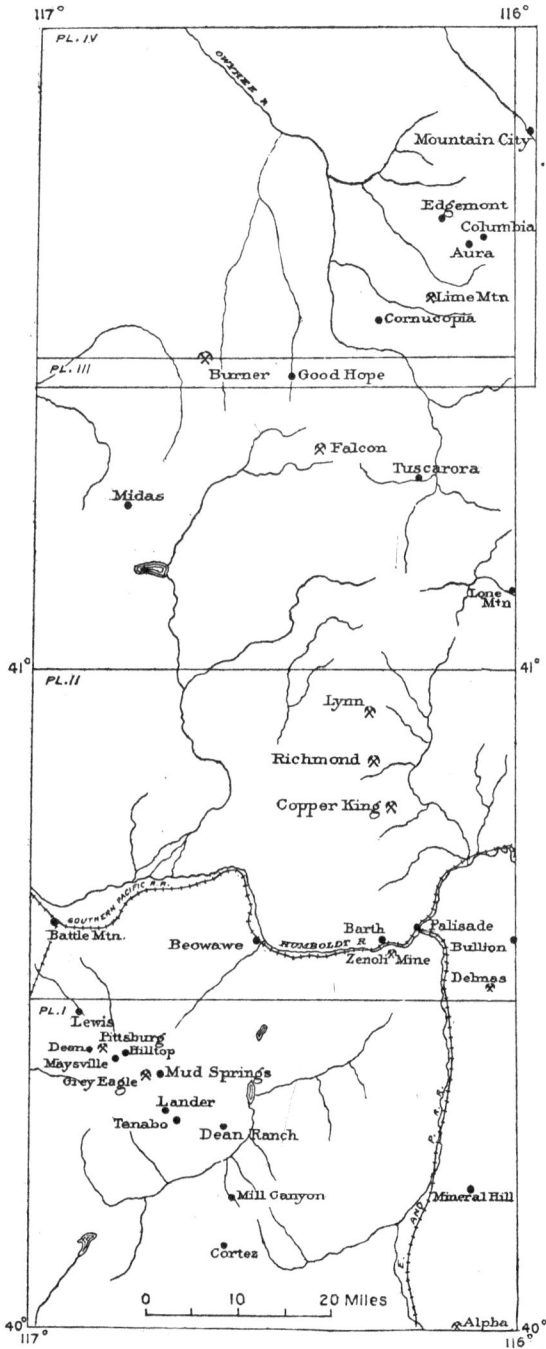

FIGURE 2.—Outline map showing areas covered by Plates I, II, III, and IV.

the valley flats produce a great variety of products, which find a good market at the ranches and mines near by. The country, especially the northern part, is an excellent range for cattle and sheep and contains some large and successful ranches.

HISTORY.

As early as 1825 the region of Humboldt River was known to British and American trappers, but settlement was slow until the discoveries of gold in California in 1849. Many of the gold seekers of that time traveled by way of the Humboldt Valley, which, during dry seasons, furnished a good roadbed, and consequently this country was well known before the more remote territory was explored. Many of the early settlers were Mormons from Utah, but after the discovery of the mines these were greatly outnumbered. The presence of silver on the Comstock lode in paying quantities became generally known in 1859, and discoveries of lode deposits were made about the same time in the Reese River country and elsewhere in Nevada. As a result of successful operations on the Comstock lode in 1860, prospecting became the principal business of a large part of the rapidly increasing population, and in the following ten years discoveries were made at nearly all the mining camps in the central and northern parts of the State.

Silver mining reached high tide in the seventies and eighties and suffered a rapid decline in the nineties. The more recent discoveries of gold and silver in the southern part of the State gave a new zest to prospecting, and in 1906 and 1907 nearly every one of the old camps was overhauled, the mines and dumps were sampled, and in some of them new ore bodies were discovered. Several new camps were established, among them Gold Circle, Lynn, Tenabo, Hilltop, and Aura. As a result of the business depression of 1907, there was a sudden cessation of prospecting and many of the mines were closed. A number of unworthy flotations were stopped and some of the more promising projects were hindered by the withdrawal of capital from the purchase of mining shares. Since money has become more plentiful, some of the prospecting operations have been taken up again. Historical data respecting the development of the various camps are given where these camps are described.

PRODUCTION.

The total production of the mining camps of this area can not be closely approximated. Official reports relating to the periods of greatest production for most of the camps were not accessible to the writer, and the government reports which are at hand do not give sufficient detail to permit more than the rudest kind of an estimate. The production of Tuscarora, the largest camp, will probably fall between $25,000,000 and $40,000,000. From the remaining camps, among which are Cortez, Mineral Hill. Bullion, Dean, Cornucopia, Edgemont, Mountain City, and others. the total is about $25,000,000 more. A conservative estimate thus gives a total production of about $50,000,000 for the entire area. On the other hand, if maximum figures are used, based largely on hearsay, the estimate of production would be considerably greater, reaching about $70,000,000. Perhaps one-quarter of the total production is gold, a large part of which is from the placers at Tuscarora and the Dexter mine and from the silver bullion from various silver-mining camps, nearly all of which produce some gold. Except for relatively small amounts of lead, copper, and zinc which have been shipped to smelters in recent years, the remaining three-fourths of the production is silver.

In 1907 the production of the area was approximately $350,000. Of this amount Edgemont, Aura, and Mountain City produced $161,834, of which $145,525 was gold, $16,004 silver, and the remainder lead and zinc. The larger portion of this production came from the Lucky Girl mine, at Edgemont. The mines at Bullion in the Railroad district, the Zenoli mine in the Safford district, and the Grey Eagle mine produced altogether $106,252, of which $2,857 was gold, $77,800 silver, $15,454 copper, $7,926 lead, and $2,195 zinc. The remainder of the production came from Mill Canyon, Cortez, Tenabo, and Lander.

GENERAL GEOLOGY.

SEDIMENTARY ROCKS.

INTRODUCTION.

The country covered by this reconnaissance is an area of Paleozoic sedimentary rocks cut by granodiorite and other intrusive rocks and overlain by Tertiary lake beds and lava flows. In this part of Nevada none of the sedimentary or igneous rocks have suffered deep-seated dynamic metamorphism such as results in the development of gneissoid or schistose structure and none of them have been deformed in the zone of flow. Around the intrusive granular rocks there is noticeable contact metamorphism, with garnet zones, especially in limestone and in shales. Because of the abundance of faults in this region it would be necessary to work with great detail to gain an adequate knowledge of the sedimentary column. Such work has been done in the Eureka district, where the Paleozoic column is best known, but as that work was undertaken after the Fortieth Parallel report had been completed, the sedimentary rocks in the country covered by the King report have not been studied in the light of the information gained at Eureka. Consequently, only broad statements can be made respecting the age of the sedimentary rocks, and on account of the small number of described exposures containing fossils much confidence should not be placed in the correlations of the rocks which have been made in a very hasty

Geologic section in Eureka region, Nevada.

System.	Formation.	Thickness (feet).	Character.
Carboniferous.	Upper coal measures........	500	Light-colored blue and drab limestones.
	Weber conglomerate.........	2,000	Coarse and fine conglomerates, with angular fragments of chert; layers of reddish-yellow sandstone.
	Lower coal measures.........	3,800	Heavy-bedded dark-blue and gray limestone, with intercalated bands of chert; argillaceous beds near the base.
	Diamond Peak quartzite....	3,000	Massive gray and brown quartzite, with brown and green shales at the summit.
	White Pine shale............	2,000	Black argillaceous shales, more or less arenaceous, with intercalations of red and reddish-brown friable sandstone, changing rapidly from one locality to another; plant impressions.
Devonian.	Nevada limestone...........	6,000	Lower strata indistinctly bedded, saccharoidal texture, gray color, passing up into strata distinctly bedded, brown, reddish brown, and gray in color, in places finely striped, producing a variegated appearance; upper rocks massive, well bedded, bluish-black in color; highly fossiliferous.
Ordovician.	Lone Mountain limestone....	1,800	Black, gritty beds at the base, passing into a light-gray siliceous rock, with all traces of bedding obliterated; Trenton fossils at the base; *Halysites* in the upper portion.
	_____Unconformity._____ Eureka quartzite............	500	Compact vitreous quartzite, white, blue, passing into reddish tints near the base; indistinct bedding.
	Pogonip limestone...........	2,700	Interstratified limestone, argillites, and arenaceous beds at the base, passing into purer, fine-grained limestone of a bluish-gray color, distinctly bedded; highly fossiliferous.
Cambrian.	Dunderberg shale *a*..........	350	Yellow argillaceous shale; layers of chert nodules throughout the bed, but more abundant near the top.
	Hamburg limestone.........	1,200	Dark-gray and granular limestone; surface weathering rough and ragged; only slight traces of bedding.
	Secret Canyon shale.........	1,600	Yellow and gray argillaceous shales, passing into shaly limestone; near the top interstratified layers of shale and thinly bedded limestones.
	Eldorado limestone *b*.........	3,050	Gray, compact limestone; lighter in color than the Hamburg limestone, traversed with thin seams of calcite; bedding planes very imperfect.
	Prospect Mountain quartzite.	1,500	Bedded brownish-white quartzites, weathering dark brown; ferruginous near the base; intercalated thin layers of arenaceous shales; beds whiter near the summit.

a This name replaces "Hamburg" shale. See Walcott, C. D., Smithsonian Misc. Coll., vol. 53, No. 1812, 1908, p. 184.
b This name replaces "Prospect Mountain" limestone. Idem.

reconnaissance. Devonian fossils have been described[a] from several places in the Pinyon Range, among them Chimney station, Hot Spring Creek, and Pinyon Pass. A fauna closely related to the Helderberg fauna has been found at White's ranch, about 20 miles north of Beowawe, and was described by Hall and Whitfield.[b] Carboniferous fossils[c] have been found at Moleen Peak, at Lone Mountain, and at Railroad Canyon, on the north end of the Diamond Range, which is about 15 miles east of Chimney station.[a]

Plate V shows the general distribution of the sedimentary rocks for part of the area. The descriptions of the sedimentary rocks in the table on page 16 are after Hague and Walcott,[d] based on the Eureka section, with modifications by E. O. Ulrich as to the age of the Pogonip and Lone Mountain limestones and the Eureka quartzite, by G. H. Girty as to the age of the White Pine shale, and by C. D. Walcott as to the "Hamburg" shale and "Prospect Mountain" limestone.

CAMBRIAN SYSTEM.

Prospect Mountain quartzite.—This formation is the earliest of the Paleozoic rocks and consists of 1,500 feet of bedded brownish-white quartzite weathering dark brown, but white near the top. The quartzites are ferruginous near the base and contain intercalated layers of arenaceous and micaceous shale only a few feet thick. The Prospect Mountain differs from the Eureka quartzite, the next overlying siliceous formation, in being more ferruginous and in general less uniform in texture, carrying throughout more or less clayey material, whereas the Eureka is a nearly pure, highly altered sandstone. No fossils were found in the Prospect Mountain quartzite.

Eldorado limestone.—The Eldorado ("Prospect Mountain") limestone overlies the Prospect Mountain quartzite and consists of 3,050 feet of gray compact limestone imperfectly bedded and of light color, which at some places is traversed by thin seams of calcite. It is difficult to define sharply the characteristic features of this formation, as changes were frequent in the deposition of the sediments, not only in vertical but also in lateral extension. In general, however, the formation has a light bluish-gray tint when observed over large areas, although nearly all colors from white to black are found in the limestone, which at the same time is characterized throughout its thickness by seams of calcite varying from one-half inch to 6 inches in width and locally forming a network of white bands. The limestone is crystalline and granular over wide areas and at many places stratification is not well shown. The *Olenellus* fauna is found in shaly beds at the base of the limestone.

---

a U. S. Geol. Expl. 40th Par., vol. 3, 1870, p. 557.
b Idem, p. 609.
c Idem, vol. 1, 1878, p. 248.
d Hague, Arnold, Mon. U. S. Geol. Survey, vol. 20, 1892, p. 13.

Secret Canyon shale.—The Eldorado limestone grades into the yellow and gray argillaceous Secret Canyon shale, which is 1,600 feet thick. Near the top are interstratified layers of shale and thinly bedded limestones.

Hamburg limestone.—The Hamburg limestone, which overlies the Secret Canyon shale, consists of 1,200 feet of dark-gray granular limestone that shows only traces of bedding. Layers of fine sandstone and hard cherty bands occur at irregular intervals.

Dunderberg shale.—The Dunderberg ("Hamburg") shale overlies the Hamburg limestone and consists of yellow argillaceous shale with layers of chert nodules throughout the bed but more abundant near the top. Across its broadest development it measures 350 feet, but it rarely maintains a uniform thickness for long distances. The Dunderberg shale carries a well-developed upper Cambrian fauna.

ORDOVICIAN SYSTEM.

Pogonip limestone.—The Dunderberg shale passes gradually into the Pogonip limestone, which is of Ordovician age. This formation is 2,700 feet thick and consists of argillites and arenaceous beds at the base, passing into purer fine-grained limestone of a bluish-gray color, distinctly bedded and highly fossiliferous. The limestone is rich in Ordovician fossils.

Eureka quartzite.—The Eureka quartzite, of Ordovician age, consists of 500 feet of compact vitreous quartzite, white or blue, passing into reddish tints near the base, with indistinct bedding. The formation is massive and very thickly bedded and has a tendency to form cliffs or escarpments. After the Eureka quartzite was deposited it was raised above sea level, but it was not greatly eroded, for in the sections studied it is everywhere present. During the subsequent subsidence it was not everywhere submerged, and so different formations will be found to overlie it at different places.

Lone Mountain limestone.—The change is everywhere abrupt between the Eureka quartzite and the Lone Mountain limestone, and the latter is not everywhere present above the Eureka and at some places is very thin. At Eureka the Lone Mountain limestone is 1,800 feet thick. The black gritty beds at the base pass into a light-gray siliceous rock without traces of bedding. Trenton fossils are found at the base and *Halysites* in the upper portion.

C. D. Walcott[a] made the following section across Lone Mountain:

Section across Lone Mountain, Nevada:

	Feet.
Dark-gray limestone, with brown and variegated layers interbedded; typical Devonian fauna (Nevada limestone)	*1,500
Siliceous bluish-gray limestone, breaking up into shaly bands carrying abundant fossils of the Lower Devonian (Nevada limestone).	200

---

a Hague, Arnold, Geology of the Eureka district, Nevada: Mon. U. S. Geol. Survey, vol. 20, 1892, pp. 61–62.

Feet.

Siliceous limestone, light brown, gray, and buff in color, with *Halysites catenulatus* near the base; passing up into beds almost white, with blue and gray tints, followed by alternating dark and light beds (Lone Mountain limestone)........................ 2,000

White quartzite (Eureka quartzite 200

Dark-gray limestone, massive bedding, with intercalated shaly layers carrying a typical Silurian [Ordovician] fauna (Pogonip limestone)... 300

Siliceous cherty limestone... 75

_____

4,275

DEVONIAN SYSTEM.

The Lone Mountain limestone passes by transition upward into the Devonian rocks. Of these the Nevada limestone is 6,000 feet thick at Eureka. The lower strata are indistinctly bedded, have a saccharoidal texture and a gray color, and pass upward into shales which are distinctly bedded, brown, reddish brown, and gray in color, and in places finely striped or variegated. The upper strata are massive, well bedded, bluish black, and highly fossiliferous. The intercalated bands of argillaceous shale and quartzite vary greatly in width but do not mark any especial part of the limestone, except that they are more abundant in the middle portion than elsewhere. The limestones are everywhere more or less magnesian and nearly pure dolomites occur at many places in narrow layers. The Modoc section [a] of the Nevada limestone is given below:

Section of Nevada limestone across ridge between Signal and Modoc peaks, Nevada.

Feet.

Dark-gray to bluish-black massive limestone poor in fossils; quite well bedded; weathering partly smooth and dark colored; partly rough and pitted and of lighter color; mostly compact and massive, also of uneven texture; with numerous calcite seams........ 1,200

Light and dark colored limestone with *Stromatopora* and *Chætetes;* contains two layers thinly bedded (fissile)..................... 150

Compact light-yellow sandstone.................................. 60

Light and dark colored limestone in layers 10 to 20 feet thick, with *Stromatopora* and *Chætetes*.................................. 240

Dark-colored limestone with *Stromatopora* and *Chætetes*......:... 180

Alternating layers (about 10 feet thick) of dark and light gray limestone, finely banded and lined; weathering brownish gray; in places bearing *Chætetes*...................................... 900

Compact yellow sandstone....................................... 50

Dark and light gray limestone; indistinct bedding................ 150

Compact yellow sandstone....................................... 50

Dark and light colored limestone interbedded in layers from 4 to 10 feet thick... 250

Light-gray siliceous limestone; very siliceous near base.......... 270

Alternating beds of dark and light gray limestone; at base 30 feet [of] very siliceous limestone; with cross-bedding on weathered surface... 180

_____

[a] Hague, Arnold, op. cit., p. 66.

	Feet.
Compact yellow sandstone	30
Dark and light gray limestone in thick belts of dark, lighter, and gray colors	575
Dark dense limestone, well bedded; bearing fossils	225
Shaly limestone, rich in fossils	200
Light-gray siliceous limestone, with fine lines of bedding; in upper portion weathering in almost rectangular fragments; growing less siliceous toward the bottom	550
Light-gray, highly crystalline, saccharoid dolomite; not siliceous	140

The Nevada limestone, like the Lone Mountain, was deposited in a sinking sea bottom, and consequently at some places it rests upon the Eureka quartzite.

CARBONIFEROUS SYSTEM.

White Pine shale.—This formation, included in the Devonian in the early reports, has for several years been regarded as early Carboniferous by G. H. Girty, who correlates it with the Caney shale of Oklahoma in Bulletin 377 of the United States Geological Survey. It consists of 2,000 feet of black, argillaceous, somewhat sandy shales, with intercalations of red and reddish-brown friable sandstones. It rests conformably upon the Nevada limestone and occupies a clearly defined stratigraphic position, with a marked change in the character of sedimentation and a fauna distinct from those of both the underlying and overlying formations. Fossil plant remains are common and all evidences indicate that the formation is a shallow-water deposit.

Diamond Peak quartzite.—This formation consists of 3,000 feet of massive gray and brown quartzite with brown and green shales at the top. At the base fine conglomerates lie next to the argillaceous White Pine shale, but a short distance up these beds pass into a more massive, usually vitreous quartzite with a characteristic gray-brown color, which breaks irregularly with a fluty fracture. A narrow belt of blue limestone occurring about 200 feet above the base of the formation carries Carboniferous fossils. This formation seems to be absent in sections described in the report of the Fortieth Parallel Survey.

Lower coal measures.—The lower coal measures consist of 3,800 feet of heavy-bedded dark-blue and gray limestones with intercalated bands of chert and with argillaceous beds near the base. The beds rest conformably upon the Diamond Peak quartzite and are very extensively developed in Utah and northern Nevada. They contain an abundant fauna representing a commingling of species from the upper and lower Carboniferous.

Weber conglomerate.—The Weber is 2,000 feet thick at Eureka and consists of coarse and fine conglomerates with angular fragments of chert and layers of reddish-yellow sandstone. At Eureka the material gives abundant evidence of shallow-water deposition and shows the existence of a land mass not very far removed at the time of deposition. The fragments are rounded pebbles of limestone, quartzite, flint, and jasperoid, evidently derived from early Paleozoic rocks. In the Shoshone Range there is a great thickness of fine-grained quartzites or siliceous shales, with conglomerates very subordinate in quantity, which are said to belong to the Weber. According to Clarence King [a] the Weber is 6,000 feet thick in the Wasatch Mountains and 8,000 feet thick in the Oquirrh Mountains.

Upper coal measures.—These beds at Eureka are 500 feet thick and consist of blue and drab limestones carrying fossils of the upper Carboniferous period. Clarence King describes the upper Carboniferous limestone as a body 2,000 feet thick, which over the Great Basin country is prevailingly made up of light-gray or drab limestone and is as a rule thinly bedded. Some carbonaceous shales were found within these measures, but no workable coal has been discovered in the Great Basin. Near the base of the upper coal measures in the Eureka district there is an intraformational conglomerate in the pebbles of which there are organic remains such as are common in rocks below the Weber conglomerate. No Permian beds are known in this region.

MESOZOIC ROCKS.

No Mesozoic sedimentary rocks are represented in the area here discussed. Throughout the Triassic and Jurassic periods either this area was a land mass in which no beds were laid down or else such beds, if they were deposited, were subsequently eroded. Triassic beds are extensively developed in the Humboldt, Pahute, and Havallah ranges, north and east of Carson Lake, and Jurassic beds occur in the Montezuma Range and in the Eugene Mountains.

TERTIARY ROCKS.

Eocene.—No exposures of Eocene rocks are known to be present in the area here described, but a few miles east of the eastern border Eocene beds with coal seams are found at several places. The western border of the Eocene was presumably east of the one hundred and sixteenth meridian. On the northwest slope of the Elko Range, about 3 or 4 miles east of Elko, the Eocene beds are exposed, dipping 35° E.[b] These beds consist of very thin shales, calcareous at some places and siliceous at others, and they contain seams of impure coal. They are overlain unconformably by the Humboldt formation (Pliocene), which here consists of white, porous, volcanic ash. In the Dixie Hills region, near the head of Dixie Valley, there are finely bedded calcareous shales and marls containing carbonaceous seams

a U. S. Geol. Expl. 40th Par., vol. 1, 1878, p. 240.
b Idem, vol. 2, 1877, p. 602.

and coal. The beds dip 30° E. and resemble those at Elko in composition. These exposures are farther west than any others of Eocene rocks. On the east front of the River Range, about 14 miles north of Osino, Eocene beds, including coal seams, dip 45° S. and are overlain by beds of volcanic ash belonging to the Humboldt formation (Pliocene).

Miocene.—No sedimentary rocks of Miocene age are known to have been deposited in the area studied, but west of it Miocene beds are extensively developed. Fossiliferous Miocene beds, which King has called the Truckee group, appear in the Kawsoh Mountains and along the south end of the Montezuma Range, and beds lithologically similar are present in the Reese River canyon. The Miocene beds contain much volcanic débris of a rhyolitic character and nearly everywhere are overlain by rhyolite.

Pliocene.—In Pliocene time a great lake occupied almost the whole territory between the Wasatch Range on the east and the Sierra Nevada on the west, extending northward far into Idaho and southward to an unknown distance. This lake Clarence King [a] has named Shoshone Lake, and the beds laid down in it are called the Humboldt formation. These beds are composed mainly of friable gray, white, and drab sandstone and marly limestones and at many places contain abundant volcanic material, chiefly tuff of a rhyolitic character. Some of the siliceous beds are made up largely of diatomaceous earth. In Bone Valley, just west of the Mallard Hills, which are some 30 miles north of Halleck, Pliocene fossils have been found.

The thickness of the Humboldt formation is at most places unknown, for complete sections are very rare. In the Huntington Valley, according to King, it can not be less than 600 or 800 feet, and occurrences at other places give the impression that the thickness is greater.

The Humboldt formation covers large areas in this part of Nevada. It is found on the east and west slopes of the Pinyon Range, on the east slope of the Cortez Range, and in large areas along Maggie Creek, in Rock Creek and Squaw valleys, and in the Owyhee Desert. At several places it lies approximately flat and rests unconformably on the upturned edges of the steeply tilted Eocene beds. In the eastern foothills of the Cortez Range, east of Cortez Peak,[b] the Humboldt beds overlie an extensive basalt flow. Here the beds are arenaceous and near the top pass into fine clays; some of them contain a large proportion of lime carbonate and sodium carbonate. The formation seems to extend over the whole of Pine and Garden valleys, but only on the outer margins of these valleys are the Quaternary gravels cut through so as to expose the Humboldt. About 5 miles north of Mineral Hill the strata, which are here highly calcareous, form ver-

---

a U. S. Geol. Expl. 40th Par., vol. 1, 1878, p. 454.
b Idem, vol. 2, 1877, p. 583.

tical bluffs 50 feet high. At this place the Pliocene was clearly eroded before the Quaternary beds were deposited.

In the Squaw Valley region there are extensive beds of the Humboldt composed largely of volcanic ash. Farther north, in the Owyhee Desert, there are very large areas of the same rocks, interbedded with basalt flows. In that region the Humboldt beds are at some places tilted, and east of the Burner Hills they form a monocline, dipping steeply away from that uplift. As a rule, however, the Humboldt formation is not so steeply tilted as earlier formations, and at several places the beds are flat-lying, resting upon the more highly tilted Eocene beds. In the Centennial Range east of the Bull Run mine, and high up on the slope south of Bull Run Creek, there are Tertiary beds which contain volcanic material of considerable size. These beds carry some thin seams of impure coal and are tilted and faulted against the Paleozoic sedimentary rocks.

QUATERNARY DEPOSITS.

The Quaternary formations comprise the alluvial deposits which nearly everywhere cover the broad valleys between the mountains and the glacial deposits which are found on the slopes of the most lofty of the mountain ranges.

Glacial deposits.—Many of the mountains show no evidence of glaciation. Their summits are rounded domes and the streams which drain them flow through sharp, narrow canyons that are barren of glacial drift, but in some of the higher ranges the evidence of glaciation is clear and unmistakable. In the higher part of the Shoshone Range to the south of Battle Mountain the amphitheater at the head of the streams could have been formed only by the action of ice. The glacial features of the range were described by Hague,[a] who records the presence of glacial débris and striæ at the foot of Shoshone Peak. The central portion of the Centennial Range is glaciated also, as is shown by the morainal material in the small canyons which trend from the summits of the range toward Aura. The Jack Creek Range (or north end of the Seetoya Range) was not visited by the writer; the amphitheaters at the heads of the canyons which drain this range indicate that it was subject to erosion by ice. All these glaciated groups rise more than 9,000 feet above sea level, but most of the mountains of this region do not reach that elevation and were not glaciated.

Alluvial deposits.—Nearly everywhere the valleys between the mountain ranges are covered by great accumulations of Quaternary gravels. These alluvial deposits were probably formed before, during, and after the glacial period, and it is not likely that they were affected by the local glaciation to any important extent.

---

[a] U. S. Geol. Expl. 40th Par., vol. 2, 1877, p. 620.

In Quaternary time—according to King, in the glacial period—two great lakes formed between the Wasatch Range on the east and the Sierra Nevada on the west. Lake Bonneville, the remnants of which are to-day represented by Great Salt Lake and Utah Lake, extended from the foot of the Wasatch Range to the Gosiute Range. Lake Lahontan, which was nearly as large, extended eastward from the region of Pyramid Lake nearly to Battle Mountain. The country herein described, except a small area along the valley of Humboldt River near Battle Mountain, which is represented on the King map [a] as occupying a long embayment of Lake Lahontan, was situated in the higher country between these two lakes. There were, however, two smaller lakes east of the Shoshone Range. One of these was in the Crescent Valley west of the Cortez Range, and another was west of Carico Peak, its remnant being now represented by Carico Lake.

The Quaternary deposits are made up of fragmental material of various sizes. At some places the fragments are well rounded and at others they are angular. All the sedimentary and igneous rocks are represented in these fragments, the proportions depending on the nearness and size of their outcrops. Some of the alluvial deposits are doubtless subaerial accumulations formed largely by freshets resulting from the frequent cloud-bursts, but others were formed under water, as is shown by their stratification and by the fact that they are rich in saline matter, which has resulted from the drying up of the inclosed basin. In the Crescent Valley, between the Cortez and Shoshone ranges,[b] the Quaternary beds include deposits of relatively pure sodium chloride which has been used for metallurgical purposes.

IGNEOUS ROCKS.

INTRUSIVE GRANITIC ROCKS.

DISTRIBUTION AND COMPOSITION.

The oldest igneous rocks in the area are stocks and dikes of granular rocks which intrude the Paleozoic sedimentary rocks at many places. These intrusive rocks, which are commonly called granite, are as a rule medium grained and of light-gray color and are composed of quartz, feldspar, and biotite, with a variable amount of hornblende. None of the granitic rocks are sheared or contorted in the manner which results when deeply buried rocks are metamorphosed. Under the microscope they show a considerable range in composition. Some of them may properly be called quartz diorites or quartz monzonites, but most of them could be classified as granodiorites, if this term is used with its broadest application. Granodiorite is, according to Lindgren,[c]

---

a U. S. Geol. Expl. 40th Par., vol. 1, 1878, Pl. VI.
b Idem, p. 503.
c Lindgren, Waldemar, Am. Jour. Sci., 3d ser., vol. 46, 1893, p. 203.

a light-gray granitic rock consisting in typical development of feldspar, quartz, biotite, and hornblende, with a medium-grained texture. The soda-lime feldspars are usually considerable and to a variable extent in excess of the alkali feldspars. The silica varies between 60 and 73 per cent; the amount of lime is variable, but it rarely exceeds and usually falls somewhat short of the alkalies, while in some varieties, which can not be distinguished from the others in the field, there is more potash than soda; a frequently occurring relation is 2 per cent K_2O to 4 per cent Na_2O. It will be seen that the rock very closely approaches some quartz-mica diorites and often might be indicated by that name.

Specimens from four of the older intrusive masses are low in potash and relatively high in soda and lime. One of these intrusive bodies is in the Railroad mining district near Bullion; another forms the summit of Lone Mountain; a third is a large body forming the mountain on which the Grey Eagle mine is located; and a fourth is near the head of Willow Creek, about 8 miles west of Tuscarora. These stocks could properly be called quartz diorites and are higher in lime and in soda than any of the other stocks. They are composed characteristically of feldspars, quartz, biotite, and a little hornblende, with magnetite and apatite as accessory minerals. The feldspars are mainly plagioclase and for the most part andesine. Only a little orthoclase is present.

Intrusive rocks of this group which approach more nearly the mean composition of granodiorite are a number of broad dikes and stocks in the Centennial Range near Edgemont and Aura and several large masses in the Shoshone Range near Dean and Tenabo. The constituent minerals are the same as in the rocks described above, but there is a little more orthoclase and quartz and plagioclase is not quite so abundant.

An analysis[a] of the rock which constitutes the summit of Shoshone Peak was made by R. W. Woodward. This analysis gives all the iron as ferrous iron and the norm can not be calculated, but it is sufficient to confirm the classification of the rock as granodiorite. Its composition is very near that of a granodiorite from Pyramid Peak, Eldorado County, Cal., described by Lindgren,[b] and that of other typical granodiorites in California.

Analysis of granodiorite from Shoshone Peak.

[By R. W. Woodward.]

SiO_2	70. 17
Al_2O_3	14. 53
FeO	4. 03
CaO	2. 29
MgO	.93
Na_2O	3. 25
K_2O	3. 35
H_2O	1. 53
	100. 08

a U. S. Geol. Expl. 40th Par., vol. 2, 1877, p. 621.
b Lindgren, Waldemar, Am. Jour. Sci., 4th ser., vol. 3, 1897, p. 306.

Still another group of the older intrusive rocks includes the stocks at Cortez and Mill Canyon, at Lone Mountain, and at Mountain City. The rocks of this group contain considerable orthoclase. Some of the specimens studied are composed of feldspar, quartz, mica, and hornblende. Orthoclase and quartz are more abundant than in the granodiorite intrusives above referred to. Plagioclase is less abundant and not so high in lime and is in the main oligoclase and andesine. A single stock may show notable differences in composition. The more acidic phases of these rocks are quartz monzonites and the more basic phases granodiorites.

An analysis made by Thomas M. Drown [a] of a specimen taken from Lone Mountain or Nannies Peak by S. F. Emmons is given below. This rock is a quartz monzonite and in composition approaches closely the quartz monzonite of the Idaho Hailey mine, Hailey, Idaho, described by Waldemar Lindgren,[b] and the quartz monzonite near San Miguel Peak, Telluride, Colo., described by Whitman Cross.[c]

Analysis of quartz monzonite of Lone Mountain (Nannies Peak).

[By Thomas M. Drown.]

SiO_2	70.77
Al_2O_3	15.22
FeO	2.65
MnO_2	.11
CaO	2.33
MgO	.71
Na_2O	3.75
K_2O	4.58
H_2O	.52
	100.64

The granitic mass in the Agate Pass region was not visited by the writer. An analysis of a specimen collected by S. F. Emmons is given below. It represents a typical soda-rich granite and is considerably higher in alkali than any of the rocks collected from the area studied.

Analysis of granite from Agate Pass.

[By Thomas M. Drown.]

SiO_2	72.01
Al_2O_3	14.75
FeO	2.35
MnO_2	.17
CaO	.79
MgO	.65
Na_2O	4.21
K_2O	4.49
H_2O	.61
	100.03

---

a U. S. Geol. Expl. 40th Par., vol. 2, 1877, p. 603.

b Twentieth Ann. Rept. U. S. Geol. Survey, pt. 2, 1900, p. 81.

c Telluride folio (No. 57), Geol. Atlas U. S., U. S. Geol. Survey, 1899, p. 6.

To the southwest of this body of soda-rich granite there is an area of quartz diorite. The mass that outcrops in Agate Pass forms the southwest half of the long area of granitic rock which is shown on the geologic map (Pl. V) on the northwest slope of Cortez Peak. As indicated in the geologic atlas of the Fortieth Parallel Survey, this quartz diorite mass is about 4 miles long and half as wide. It is composed of plagioclase, quartz, hornblende, and biotite, and the subjoined chemical analysis indicates that it is the most basic of all the granitic rocks of this group. As no ore deposits were reported in this area, it was not visited by the writer, but to judge from the description by Zirkel [a] it is highly sericitized along Agate Canyon and, as is well known, ore deposition very commonly accompanies sericitization. In composition the rock closely resembles a quartz diorite from Spanish Peak, Plumas County, Cal., described by H. W. Turner,[b] and a similar rock from Yaqui Creek, Mariposa County, Cal., described by the same writer.[c]

Analysis of quartz diorite from Agate Pass.

[By R. W. Woodward.]

SiO_2	58.54
Al_2O_3	16.68
FeO	5.62
CaO	6.00
MgO	5.22
Na_2O	2.76
K_2O	2.50
H_2O	2.15
	99.47

CONTACT RELATIONS AND AGE.

In the work of the Fortieth Parallel Survey some of the stocks were supposed to be of Archean age. The significance of contact phenomena was, however, fully recognized, for King,[d] speaking of their relations to the sedimentary rocks, says:

The configuration of the granite topography of the Archean surface prior to the deposition of the Paleozoic series was that of an area of mountain ranges, possessing some very abrupt precipitous walls, sharp lofty peaks, and broad low domes. Where these came to be uptilted together with superjacent strata, and afterward exhumed by erosion, which brought to light granite peaks piercing through highly inclined beds, it often becomes absolutely impossible to determine the relation of the two. In the absence of any granitic dikes penetrating the stratified series, or of peculiar local metamorphism, or of general evidence of intrusion, the bodies are usually referred to the old Archean topography. Only in cases where the granite is actually seen to penetrate either fissures or warped openings in the strata is it safe to refer it to a later origin than the sedimentary series.

---

a U. S. Geol. Expl. 40th Par., vol. 2, 1877, p. 576.

b Seventeenth Ann. Rept. U. S. Geol. Survey, pt. 1, 1896, p. 72.

c Bull. U. S. Geol. Survey No. 150, 1898, p. 342.

d King, Clarence, U. S. Geol. Expl. 40th Par., vol. 1, 1878, pp. 77, 100.

This statement of the criteria of intrusion is brief and comprehensive. It remains, therefore, to discuss the evidence of intrusion in some detail for each granitic mass. The more important intrusive bodies studied are located near Bullion, in the Railroad district, in the Cortez and Mill Canyon mining district, at several places in the Shoshone Range, at Lone Mountain, in the Centennial Range, and at Mountain City.

The stock near Bullion shows a very irregular contact and dikes of the granodiorite cut the limestone. The contact between the granodiorite and limestone is marked by a garnet zone several hundred feet wide.

At Cortez the relations between granodiorite and limestone are clearly crosscutting. At Mill Canyon the granodiorite stock sends out small apophyses into limestones, and near the contact the sedimentary rocks carry contact-metamorphic silicates.

In the Shoshone Range granitic rocks or their porphyries were noted at Tenabo, Mud Springs, Hilltop, Grey Eagle, and Dean. At Tenabo some veinlets of actinolite cut the quartzite near the contact with granodiorite porphyry, but no garnet zones were observed in this range. The contacts of the igneous and sedimentary rocks are distinctly crosscutting at Mud Springs, Hilltop, and Dean.

At Lone Mountain zones of contact-metamorphic silicates are developed near the contact of granodiorite and the sedimentary rocks.

In the Centennial Range the Paleozoic sedimentary rocks are cut by dikes and other intrusives of granodiorite which are clearly of later age.

At Mountain City the granitic rocks are clearly later than the sedimentary rocks. At California Hill garnet and other contact-metamorphic silicates are developed in limestone along the contact.

The contact relations of the granodiorite about 8 miles west of Tuscarora, near the head of Williams Creek, were not studied, but the rock has the same general composition as that of the stocks mentioned, and, like them, it does not show deep-seated metamorphism. In view of the evidence shown by the contacts of practically all the masses, it may be assumed that they are, without exception, intruded into the Paleozoic rocks and that they are therefore of post-Carboniferous age. They are known also to be older than the rhyolites and basalts and associated andesites, for they do not intrude these rocks. From evidence within the area studied, these stocks are therefore known to be younger than the Carboniferous and older than the Miocene, but on the assumption that the intrusives of this character throughout the West generally were formed at about the same time, the ages of these rocks may be estimated more closely. In the Humboldt Range F. L. Ransome found granodiorite stocks cutting the Triassic sedimentary rocks. In California similar stocks cut the Jurassic. At Bisbee, Ariz., as shown by Ransome, granitic stocks cut the Carboniferous and are eroded and covered by sediments of Comanche (Lower Cretaceous) age—the "Bisbee group" of Ransome's report.[a] In the Clifton-Morenci district, according to Lindgren,[b] the granite porphyry and related deep-seated porphyries cut Cretaceous beds which are equivalent to the Benton, and they are therefore probably of late Cretaceous or the very earliest Tertiary age. At Ely, Nev., at Bingham, Utah, and at other Utah camps deep-seated granular rocks intrude sedimentary rocks, but at those places no sedimentary rocks later than the Carboniferous are represented and the age of the intrusives can not be closely approximated. Near Philipsburg, Mont., large stocks of granites and nearly related rocks cut the Paleozoic and Mesozoic sediments, including beds as young as the Cretaceous. From the relations thus shown by the post-Paleozoic granular or porphyritic intrusive rocks elsewhere in the Western States it may with some reason be assumed that these intrusives are of about the same age, and that they were formed either at the beginning of the Cretaceous, during the Cretaceous, at the very end of Cretaceous time, or perhaps at the very beginning of the Tertiary.

PORPHYRIES ASSOCIATED WITH INTRUSIVE GRANITIC ROCKS.

Porphyritic dikes and other small intrusive masses are associated with some of the granitic stocks. Most of these are believed to be either the more rapidly cooled portions or the products of differentiation from the granodiorite magmas. In the main they were formed at about the same time that the stocks were intruded, but some of them may be younger and may belong to the late Tertiary period of volcanism.

An intruding quartz porphyry in the Standing Elk mine, near Bullion, contains abundant rounded phenocrysts of quartz and a smaller number of feldspar phenocrysts in a light-colored, fine-grained groundmass. Muscovite and biotite are present, but the dark-colored silicates are very sparingly developed and are much less abundant than in the granodiorite with which the porphyry is associated and from which it is probably derived.

At Lone Mountain granodiorite porphyry is associated with granodiorite and appears to grade into it. The granodiorite porphyry has a fine crystalline groundmass composed of quartz and orthoclase, in which there are phenocrysts of acidic plagioclase, resorbed quartz, biotite, and hornblende. Orthoclase and quartz are more abundant in the porphyry than in the granodiorite.

---

a Ransome, F. L., Prof. Paper U. S. Geol. Survey No. 21, 1904, pp. 56-73.
b Lindgren, Waldemar, Prof. Paper U. S. Geol. Survey No. 43, 1905, p. 85.

The granodiorite of the Phoenix mine, at Tenabo, is cut by a quartz porphyry, the microcrystalline groundmass of which contains abundant phenocrysts of orthoclase, quartz, and biotite, with little or no plagioclase. A similar porphyry cuts the sedimentary rocks in the Gold Quartz mine, near by. At the Two Widows mine, in the same camp, a small dike of quartz diorite porphyry cuts the siliceous sedimentary rocks. In the Maysville district and at Dean, on the north slope of the Shoshone Range, granodiorite porphyries cut the sedimentary rocks. These have a coarsely microcrystalline groundmass and grade into granodiorite.

The granodiorite at Mountain City is cut by small, light-colored dikes of fine-grained granite, some of which are typical aplite. They are probably differentiation products of the granodiorite magma.

At Cortez and at Mineral Hill several dikes cut the limestone. These are composed mainly of quartz, sericite, calcite, and iron oxide and are too much decomposed for determination. They are probably derived from the deep-seated intrusives.

A felsitic dike of quartz porphyry on the west slope of Lone Mountain and an intrusive of somewhat similar character below the Pittsburg mine, at Dean, are believed to belong to a period of volcanism later than that during which the granodiorites and related porphyries were formed. The groundmass of these quartz porphyries is much the same as that of rhyolite, indicating that they cooled near the surface. Quartz porphyries with rhyolites and andesites are discussed on another page.

TERTIARY ERUPTIVE ROCKS.

RHYOLITE.

Rhyolites are glassy igneous rocks which have about the same chemical composition as granites. In color they are white, pink, purple, or dark brown. The dense pasty groundmass may contain phenocrysts of quartz and orthoclase, with small amounts of biotite, augite, and hornblende. A little soda-rich plagioclase may also be present among the phenocrysts. They form surface flows and many of them show streaking due to flowage. Some of them are flow breccias formed of angular fragments of rhyolite in a matrix of the same. Such rocks result when a crust forms over the flow and is broken up by the movement of the still liquid portion, which, solidifying, forms the matrix for the fragments. Some of the rhyolites in the Gold Circle district are thin, fissile bands which resemble shales. The shaly appearance is probably due to banding that developed as the rhyolite flowed and was emphasized by subsequent weathering along the parting planes. Some of the rhyolites are vesicular. The small blebby holes in these rocks represent the places where imprisoned gases expanded when the pressure was removed from the magma at the time of eruption. In some of the rhyolites in the Gold Circle district the vesicles are filled with amygdules of amethystine quartz, deposited by water after the rock had solidified.

The term rhyolite in this reconnaissance is used in a broad sense. Some of the rocks so called are probably latites. Some of the brown glasses which have been called obsidians contain, besides quartz phenocrysts, acidic plagioclase, augite, and some magnetite and in composition approach quartz-bearing andesites.

The rhyolites are the most abundant and most widely distributed of the Tertiary eruptive rocks; there is not an important mountain range which does not contain large bodies of them, but they are most numerous in the northern and western parts of the area. They are the oldest Tertiary eruptives and at many places rest directly on the Paleozoic sedimentary rocks. They are cut by andesites and overlain by flows of basalt. Their thickness has not been measured but must reach a maximum of over 2,000 feet. From many places they have probably been eroded. Their distribution in the various mountain ranges is discussed in greater detail where the rocks of the various ranges are described.

BASALT.

Basalts are dark basic igneous rocks with a dense compact ground-mass which may be finely microcrystalline, but which nearly every-where contains dark-colored glass. Basic plagioclase, olivine, pyroxene, and magnetite are generally present, some of them as crystals visible to the eye. The darker color and the presence of the small crystals of greenish-yellow, glasslike olivine will usually serve to distinguish the basalt from andesites. Some of the basalts are dense and glassy and others are highly vesicular. In this area the basalts are not so widely distributed as the rhyolites, but there are several bodies of considerable size. One of these caps the great flat-topped Shoshone Mesa northeast of Battle Mountain and another covers Whirlwind Mesa west of Beowawe. A broad belt of basalt flanks the east slope of the Cortez Range for a distance of about 30 miles.

The basalts occur as flows which overlie the rhyolites and the older rocks, and as dikes which cut them. At some places they are over-lain by Pliocene lake beds and at other places they rest upon these beds. The thickness of the basalt flows has not been determined at many places. The mass which caps Shoshone Mesa is, according to King, more than 1,000 feet thick. They are not known to have been intruded by the andesites nor have any ore deposits been found in them. They were presumably extravasated after the period of andesitic eruption and late Tertiary ore deposition.

ANDESITE.

Andesites are brown, gray, or greenish-gray rocks which are intermediate in composition between rhyolites and basalts. The groundmass is very finely crystalline or glassy. Andesites may be aphanitic, or without visible crystals, but as a rule they have phenocrysts of feldspar with a variable amount of augite, hornblende, and biotite. If the phenocrysts are conspicuous the rock may be termed andesite porphry. Magnetite crystals and dots of magnetite are usually numerous in the groundmass and they occur as inclusions in the phenocrysts. In the andesites of this area nearly all the feldspar phenocrysts are plagioclase, but a little orthoclase is present in some. Andesine is the most abundant feldspar of the plagioclase group; oligoclase is present in some specimens and labradorite in the darker, more basic types. Some andesites carry small rounded phenocrysts of quartz, but where quartz is present in considerable quantity the rocks are called dacite. Although some of the andesites of this area show vesicular phases, this character is not so frequently developed as in the rhyolites and basalts. At Gold Circle, Tuscarora, Cornucopia, and Good Hope the andesites are associated with rhyolite. At Tenabo, Lander, and Lime Mountain they cut through the Paleozoic sedimentary rocks or form flows above them.

The andesite at Tuscarora has a dense greenish groundmass containing highly altered phenocrysts of andesine, hornblende, biotite, and magnetite, with some orthoclase. In an open cut at the Dexter mine the relations of the andesite and rhyolite are clearly crosscutting, and as the rhyolite is a flow, the andesite must have been intruded into it. At Cornucopia the andesite intrudes rhyolites and is a dark porphyry composed of a glassy groundmass which contains many phenocrysts of basic andesine and pyroxene, with large chloritic patches that seem to have resulted from the decomposition of hornblende.

At Gold Circle the andesite is relatively rich in augite and magnetite. Andesine predominates, but some of the plagioclase is labradorite. The andesite is mainly younger than the rhyolite, for dikes of andesite cut the rhyolite and flows of andesite which are presumably connected with the dikes overlie the rhyolite flows. At one place in this district a rhyolite flow caps the andesite and an intrusive mass of quartz porphyry which has the same composition as rhyolite was noted cutting andesite. It appears that the eruptions of the rhyolite took place in the main before the andesite was formed, but that some rhyolite eruptions followed those of andesite.

The contact relations between the andesites and the basalts have not been determined. No andesitic intrusions have been noted in the Pliocene lake beds (Humboldt formation) and the andesitic eruptions are therefore probably older than these beds. As the basalt effusive rocks are closely associated with these beds, it is highly probable that the andesitic eruptions ceased before the basalts were extravasated.

Andesites cut the sedimentary rocks in the Shoshone Range and are exposed in the Gem mine, at Tenabo, and in the Lovie mine, at Lander. They have a dense glassy groundmass and at some places are vesicular, and so it is assumed that they represent flows or intrusives formed near the surface. They are dark dense rocks with an altered glassy groundmass containing phenocrysts of acidic plagioclase and some orthoclase.

An analysis of a specimen of somewhat altered andesite, collected by S. F. Emmons between Wagon Canyon and Palisade, in the Cortez Range, is given below.[a]

Analysis of andesite near summit of Cortez Range, between Wagon Canyon and Palisade.

[By R. W. Woodward.]

SiO_2	61.64
Al_2O_3	17.44
Fe_2O_3	.82
FeO	3.99
CaO	5.86
MgO	3.05
Na_2O	3.45
K_2O	1.15
H_2O	2.64
	100.04

DACITE.

When andesite contains a few phenocrysts of quartz it is termed quartz-bearing andesite, but if there is a considerable proportion of quartz—as much as 3 or 4 per cent of the volume—then the rock is dacite. The groundmass of dacite is glassy or fine grained, and the phenocrysts commonly present are feldspar, quartz, hornblende, biotite, and augite. The feldspars are mainly plagioclase and may be a little richer in soda than the plagioclases of andesite. The andesite at the Zenoli mine in the Safford district, near Palisade, is richer in quartz than the other andesites which form the country rock for the ore deposits. Northeast of this region there is a body of typical dacite containing a noticeable proportion of quartz. Another body of dacite is in the Cortez Range and extends southward from Palisade a distance of several miles. The dacites are closely related to andesites in composition and occurrence and are believed to have been erupted at about the same time.

SUCCESSION OF THE TERTIARY ERUPTIVE ROCKS.

The order of eruption or the succession of Tertiary igneous rocks in the Great Basin, as given by Baron von Richthofen,[b] is (1) pro-

---

[a] U. S. Geol. Expl. 40th Par., vol. 2, 1877, p. 587.

[b] Richthofen, F. von, Principles of the natural system of volcanic rocks: Mem. California Acad. Sci., vol. 1, 1868, p. 36; also Jahrb. K. k. geol. Reichsanstalt, vol. 11.

pylite, (2) andesite, (3) trachyte, (4) rhyolite, (5) basalt. This sequence in an amplified form was accepted by the geologists of the Fortieth Parallel Survey, but for rhyolite and basalt the term neolite was substituted. Subsequently, it was proved by G. F. Becker [a] that propylite is in the Washoe district an altered phase of andesite. Later still this form of alteration was shown to be common and so the term propylite has ceased to be used for the primary rock, but "propylitization" has been accepted for a certain kind of hydrothermal metamorphism. The trachytes, as determined by Zirkel for the Fortieth Parallel Survey, have been shown by Hague and Iddings [b] to be in the main andesites and dacites. Spurr,[c] who has studied the Tertiary lavas throughout a large part of the Great Basin, gives this general succession: (1) Rhyolite, (2) andesite, (3) rhyolite, (4) andesite, and (5) basalt.

In the area here considered the rhyolites were the first rocks erupted. They were the most extensive, for they cover a large part of the surface and their thickness is very great. Subsequently the andesites were erupted, but in much smaller amounts. They took the form of dikes and large intrusive masses as well as flows, and in that they differed from the rhyolites, which are flows. The eruption of the basalts followed that of the andesites and seems to have closed the period of volcanism in northern Nevada. The Pliocene lake beds (Humboldt formation), whose deposition followed the bulk of the rhyolite eruptions, were formed during the eruptions of the basalts, for basalt flows occur under them and interbedded with them. Some of the rhyolite appears also to have been erupted during the deposition of the Humboldt beds, for rhyolite pumice is interbedded with them, and this does not seem to be a sand washed from older rhyolites, but volcanic dust which settled directly in water. The periods of the extravasation of the various lavas are thus shown to have been to some extent overlapping, but the general succession is (1) rhyolite, (2) andesite, (3) basalt.

AGE OF THE TERTIARY ERUPTIVE ROCKS.

In Eocene time the area covered by this reconnaissance was, as already stated, dry land, but just east of it there was a great lake in which the Eocene beds were deposited. In the Dixie Hills, east of the Pinyon Range, some 25 miles southwest of Elko, these beds, according to S. F. Emmons,[d] are for the most part finely bedded calcareous shales containing carbonaceous members which carry seams of impure coal. Similar beds are found in Penn Canyon,[e] 14 miles north of

a Geology of the Comstock lode and the Washoe district: Mon. U. S. Geol. Survey, vol. 3, 1882, p. 88.
b Am. Jour. Sci., 3d ser., vol. 27, 1884, p. 453.
c Jour. Geology, vol. 8, 1900, p. 621.
d U. S. Geol. Expl. 40th Par., vol. 2, 1877, p. 562.
e Idem, p. 595.

Osino, and on the northwest slope of the Elko Range,[a] 4 miles east of Elko. There is no record of volcanic ejectamenta interstratified with these beds, and it is inferred that the country to the west of this lake was free from volcanism throughout the Eocene.

The area here discussed was dry land also during Miocene time, when there was a large lake to the west of it in which the Truckee formation was deposited. As already stated, typical exposures of these beds with fresh-water mollusks are found in the Kawsoh Mountains and along the south end of the Montezuma Range, and beds of similar lithologic composition occur in the Reese River canyon about 12 miles southwest of the southwest corner of the area.[b] These beds contain much volcanic material of rhyolitic character and nearly everywhere are overlain by rhyolite. Wherever observed in connection with basaltic eruptions they are cut through or overlain by the basalt.

In Pliocene time a lake occupied almost the whole territory between the Wasatch Range and the Sierra. In this lake were deposited the Humboldt beds, which contain abundant volcanic material, mainly of a rhyolitic character. These beds rest above basalts and flows of basalt are included in them.

The general succession of the eruptive rocks was, as already stated, (1) rhyolite, (2) andesite, (3) basalt. The rhyolite eruptions probably began in early Miocene time soon after the deposition of the Eocene lake beds, and continued through the Pliocene, for there are rhyolitic tuffs and lapilli in these lake beds. The andesites were intruded after there had been very extensive rhyolite eruptions, for great thicknesses of rhyolite are cut by andesite. As few if any andesites intrude the Pliocene lake beds, it is believed that the andesites were erupted mainly, if not altogether, late in Miocene time, probably at its close. The basalts were erupted during Pliocene time, for they are interbedded with the Pliocene lake beds.

DEFORMATION OF THE ROCKS.

The sedimentary rocks and lava flows are in few places found in the horizontal position, but nearly everywhere are tilted and faulted. In all the ranges where they are exposed the Paleozoic rocks are thrown into broad open folds and have dips which for the most part range from 15° to 45°. As shown by the atlas of the Fortieth Parallel Survey, the Cortez and Pinyon ranges are in the main anticlinal, the Seetoya or Jack Creek Range is synclinal, and the Shoshone Range is an eastward-dipping monocline modified by profound faulting. The Centennial Range is a northward-dipping faulted monocline modified by subordinate anticlines and synclines.

---

a U. S. Geol. Expl. 40th Par., vol. 2, 1877, p. 601.
b Idem, vol. 1, 1878, p. 412.

The Tertiary lava flows are at no place closely folded. The contorted bands of some of the shalelike rhyolites suggest that these rocks have been deformed by compression, but a close inspection shows that their convolutions are the result of movements which took place before the rocks had completely cooled, and are not caused by subsequent deformation. The lava beds are nearly everywhere in a tilted position, however, and at many places they are faulted. The tilting movements were evidently accomplished after the Paleozoic sedimentary rocks had been folded, and the lavas do not show so large an element of horizontal compression as the sediments.

The present attitude of the Paleozoic sedimentary rocks is doubtless due to various processes which operated at different times. As already stated, there are in the area no bedded rocks which were laid down between the close of the Carboniferous and the beginning of the Miocene, and consequently there is no evidence within the area itself which will show the time at which the deforming movements took place. Information respecting what happened during this time must therefore be gained outside of the area, at places where rocks of suitable age are known.

In California there is a general unconformity between the folded Jurassic beds and the Cretaceous beds, which are not so greatly folded. The time of the folding is therefore known to be at the end of the Jurassic. Folded Jurassic beds are found in northern Nevada as far east as the Pahute Range, and, although these beds are not covered by the Cretaceous, they were probably deformed at the time of the extensive mountain-making movements in California. It is not known whether this movement extended eastward across the Great Basin as far as the Wasatch Mountains, but at any rate the Wasatch and Uinta mountains and the country extending far to the east of them were strongly uplifted at the close of Cretaceous time. The effect of the movement at this time was probably felt also in northern Nevada.

A few miles east of Elko[a] the Eocene beds are highly tilted and overlain by volcanic materials of Pliocene age. In the Dixie Hills, southwest of Elko, and in Penn Canyon, 14 miles north of Osino, beds of the same age are highly tilted. The Miocene lavas are not so greatly deformed, and so it is supposed that the movement took place before Miocene time, or near the close of the Eocene. The mountain-making movements which are recorded in the folded Paleozoic sedimentary rocks seem therefore to have taken place at the end of the Jurassic, at the end of the Cretaceous, and at the end of the Eocene, but the extent and relative importance of each can not be shown.

---

a U. S. Geol. Expl. 40th Par., vol. 2, 1877, p. 595.

A period of deformation followed the eruption of the Miocene lavas, for these rocks at many places are highly tilted and faulted. This movement seems to have taken place without the strong compressional stresses which produced the pre-Miocene folds, and practically all the faults are of the normal type, in which the hanging wall appears to have dropped. Some of the mountain ranges were doubtless elevated at this time. Blocks of the Paleozoic sedimentary rocks which had been deformed by folding in pre-Miocene time were left in relatively exalted positions when other blocks sank away from them and formed the valleys. Since that time profound erosion has obliterated the fault scarps, but has not everywhere obliterated the entire effects of faulting, for the elevated block is still at a higher altitude than the depressed block.

A period of less intense crustal deformation followed the deposition of the Pliocene lake beds, which are in places gently warped and faulted.

RÉSUMÉ OF GEOLOGIC HISTORY.

In the area studied there is no record of pre-Cambrian events. The oldest rocks exposed are the quartzites and grits which outcrop on the crest of the Pinyon Range. Throughout Cambrian and early and middle Ordovician time this part of Nevada was the floor of a sea upon which quartzites, limestone, and shales were deposited, the whole series of pre-Silurian rocks having a thickness of more than 10,000 feet. Near the middle of the Ordovician period the sea became shallower and the Eureka quartzite was deposited. Subsequently the sea bottom was lifted above the water level without tilting the beds. The land mass did not remain long above the water, but sank slowly and gradually through late Ordovician time,[a] and was again elevated at the close of the Ordovician, and throughout the Silurian remained above sea level. In Devonian time approximately 6,000 feet of limestone, sandstone, and shale were deposited. Sedimentation was uninterrupted between the Devonian and the Carboniferous, but in the early Carboniferous there was a shallowing of the waters, and at some places as much as 3,000 feet of sandstone was deposited. This was followed by the deposition of 3,800 feet of limestone and shales, above which was laid down the Weber conglomerate, having a thickness of 2,000 to 6,000 feet. This was followed by the deposition of the upper Carboniferous limestone, up to 2,000 feet thick.

During the Paleozoic era there were thus deposited between 30,000 and 40,000 feet of sedimentary rocks. These beds carried a considerable proportion of conglomerates, sandstones, and shales and must have been deposited, in part at least, in relatively shallow water and not far from the shore. According to Clarence King,[b] the Paleozoic

---

a Hague, Arnold, Mon. U. S. Geol. Survey, vol. 20, 1892, p. 57.

b U. S. Geol. Expl. 40th Par., vol. 1, 1878, p. 247.

sea was east of a land mass whose shore was in northern Nevada near longitude 117° 30'. West of that meridian and north of the fortieth parallel there are no Paleozoic rocks in northern Nevada, but pre-Cambrian or post-Paleozoic rocks instead. As already stated, the sea bottom did not remain stationary, but oscillated from time to time and permitted a very great thickness of beds to be deposited. At the close of the Ordovician period parts of it rose and were not submerged again until the beginning of the Devonian. The close of Paleozoic time was marked by profound but relatively gentle conti-nental movements. In the vicinity of the Havallah Range, about 117° 30' west longitude, there was, according to King,[a] a pivotal line. The area to the east of this line remained submerged in Paleozoic time, and that to the west was above sea level. In Mesozoic time the conditions were reversed—the country to the east became dry land and that to the west sank below the level of the sea and received thick contributions of Triassic and Jurassic sediments. According to King, [b] immediately after the deposition of the Jurassic sediments they were folded with much horizontal compression, producing great north-south mountain ranges. The westernmost of these ranges was the Sierra Nevada, and in the Great Basin ranges were formed probably at the same time. The reason for this conclusion is that many of the mountain ranges parallel to the mountains in the area studied, but west of it, are composed of folded Jurassic rocks. The folding at this time was most intense in the Sierra Nevada and decreased eastward toward the Wasatch Range. At about this time, or else near the close of the Cretaceous, the granitic rocks, in the main granodiorites, were intruded in the Paleozoic beds. Many of the ore deposits were formed at the time of this intrusion.

With the advent of Tertiary time began a period marked by exten-sive inland lakes. The earliest of these in this region was the Gosiute Lake, in which were laid down the Eocene beds. Its western shore extended nearly to the eastern border of the area studied, this area constituting a land mass from which the sediments were derived. In this lake were deposited up to 2,000 feet of shales, clays, and lime-stones, with some beds of impure coal. The deposition of these beds was followed by mountain-making movements, in which they were tilted at some places as much as 45°. In Miocene time a great lake was formed west of the area. King has called this the Pahute Lake and the beds are called the Truckee formation. They contain much volcanic débris and record a time of great volcanic activity. This period of igneous activity differed from that of the post-Jurassic, as the rocks recording it are not granular rocks but are mainly lava flows and andesites, formed relatively near the surface. The rhyolites were first extravasated, and later these were intruded by andesites.

---

[a] U. S. Geol. Expl. 40th Par., vol. 1, 1878, pp. 731, 759.   [b] Idem, p. 733.

Many of the ore deposits were formed in connection with the andesitic intrusions.

In Pliocene time, when volcanism was still at its height, an extensive lake was formed which covered nearly the whole of the Great Basin. King [a] has called this Shoshone Lake and the beds that were laid down in it the Humboldt formation. In the lake, which may have contained lofty islands, a great thickness of sandstones, clays, and calcareous and diatomaceous shales was deposited, with a mass of volcanic material, principally of a rhyolitic character, which was blown out of volcanic vents and settled in the water of the lake. Extensive flows of basalt occurred at about this time.

The late Tertiary was marked by very extensive normal faulting, a large part of which took place after the lava flows were extravasated and after the second period of ore deposition. Many of the old mountain ranges were probably raised, and new ones may have been formed. In the Quaternary, mountain glaciers formed in the higher ranges during the glacial period. Small lakes occupied some of the depressions between the mountains, and extensive accumulations of débris, eroded from the mountains, filled the valleys.

ORE DEPOSITS.

GENERAL STATEMENT.

It has been shown that the igneous rocks of the area studied belong to two distinct periods of volcanism. The first of these was probably in Cretaceous time, and the rocks formed during that period are intrusive granular rocks and deep-seated porphyries, in the main granodiorites and granodiorite porphyries. All these rocks were formed at considerable depths, and since they were intruded this country has been greatly eroded and the capping which must have covered them when they solidified has been removed. These early intrusive rocks do not cut the Tertiary lake beds or lavas, but are confined to the Paleozoic sedimentary formations.

A large number of the ore deposits are in or near the early intrusives. These include deposits at Bullion (Railroad district), Lone Mountain, Edgemont, Columbia, Aura, Mountain City, Cortez, Mill Canyon, Grey Eagle, Dean, and Lewis, and some of the deposits at Tenabo. The deposits at Mineral Hill may also belong to this group.

The Tertiary lavas and associated rocks are in the main rhyolites, andesites, and basalts. These rocks are younger than the granodiorites and associated rocks and, for reasons given on pages 34–35, are thought to be of Miocene and Pliocene age. The ore bodies associated with the later eruptives include the deposits at Tuscarora, Cornucopia, Good Hope, Burner, Falcon, Stafford, Lynn, and Gold Circle, and probably some of those at Tenabo and Lander.

---

[a] U. S. Geol. Expl. 40th Par., vol. 1, 1878, p. 456.

With respect to metal content, the deposits of both groups carry silver, gold, copper, and lead. Silver and gold are the most important metals of both groups, but the proportion of gold to silver is greater in the earlier deposits than in those of the later group. The deposits at Edgemont and at Dean belong to the earlier group, and at these places the only metal won in important quantity is gold. The copper and lead deposits are associated mainly with the older intrusive rocks. These metals occur also with the silver and gold ores of the later group, but in smaller proportion.

EARLIER DEPOSITS (CRETACEOUS ?).

GENERAL STATEMENT.

The deposits associated with the older intrusive rocks are conservatively estimated to have produced $22,000,000. The minerals of these deposits are given in the list below:

Actinolite	Chalcopyrite.	Hornblende.	Pyrrhotite.
Apatite.	Chlorite.	Kaolin.	Quartz.
Argentite.	Chrysocolla.	Limonite.	Sericite.
Arsenopyrite.	Copper.	Magnetite.	Silver.
Azurite.	Cuprite.	Malachite.	Specularite.
Barite.	Diopside.	Manganite.	Stephanite.
Biotite.	Enargite.	Molybdenite.	Stibnite.
Bismuthinite ?	Epidote.	Muscovite.	Stromeyerite.
Bornite.	Freibergite.	Polybasite.	Tetrahedrite.
Bromyrite ?	Fluorite.	Proustite.	Tremolite.
Calcite.	Galena.	Pyrargyrite.	Zinc blende.
Cerusite.	Garnet.	Pyrite.	Zoisite.
Cerargyrite.	Gold.	Pyrolusite.	
Chalcanthite.	Gypsum.	Pyroxene.	
Chalcocite.	Hematite.	Pyromorphite.	

With respect to the intrusive rocks, the Cretaceous (?) ore deposits show various relations. The gold-bearing fissure veins at Edgemont and Bull Run are more than a mile away from the nearest granodiorite intrusions, but most of the deposits, especially those in limestone, are but a few rods away from the deep-seated intrusive rocks, or else they are associated with dikes which are probably connected with the larger intrusive bodies. The deposits of this group are (1) contact-metamorphic deposits, (2) irregular replacement deposits or chambers in limestone, (3) replacement veins and sheeted zones in limestones and shales, (4) fissure veins in quartzites, and (5) fissure veins in igneous rocks.

CONTACT-METAMORPHIC DEPOSITS.

The contact-metamorphic deposits are represented at Bullion (Railroad district), Lone Mountain, Lime Mountain, and Cortez. In point of production these deposits are not so important as other deposits which are usually associated with them, but with respect to genesis they form a distinct and interesting type. At Bullion the contact-metamorphic ore consists of garnet, calcite, actinolite, tremolite, epidote, quartz, pyroxene. pyrite, chalcopyrite, bornite, galena, and zinc blende. The metals are copper, silver, and lead. All the minerals are intergrown and were formed at the same time. They are in limestone not more than a few rods from the contact with the igneous rock and were formed by gaseous solutions from that rock at the time of the intrusion. At Lone Mountain contact-metamorphic deposits occur near granodiorite but are not extensively developed. The ore consists of calcite, garnet, actinolite, magnetite, pyrite, chalcopyrite, and other minerals and carries values in copper and silver. At Lime Mountain some ore composed of calcite, white and black mica, tremolite, and copper-bearing sulphides is presumably of contact-metamorphic origin. The deposits of the Garrison mine, at Cortez, are in the main chambers of siliceous ore in limestone, but a small amount of contact-metamorphic ore is found near a dike of decomposed porphyry. This ore consists of calcite, tremolite, actinolite, quartz, and sericite, intergrown with which is a small amount of pyrite.

The contact-metamorphic silicates are developed at Mill Canyon, in the Cortez Range, and also at Mountain City, but no contact-metamorphic ore has been found at these places. So far as known there are no contact-metamorphic zones in the Shoshone Range along the border of the intrusive granodiorite, which nearly everywhere breaks through quartzite, a rock that is not favorable for contact metamorphism. Why the granodiorites of the central part of the Centennial Range did not cause contact metamorphism is not easily understood, for all the conditions so far as known seem to be similar to those prevailing where contact metamorphism has taken place.

IRREGULAR REPLACEMENT DEPOSITS IN LIMESTONE.

The irregular replacement deposits or chambers in limestone are among the most important ore bodies in the area studied. They include the silver-lead deposits at Bullion, the silver deposits at Cortez and Mineral Hill, some of those at Mill Canyon, and probably some of the inaccessible deposits at Lewis and Lone Mountain. These deposits are found in areas which are intruded by the granitic rocks or the related porphyries, and for the most part they have the form of chimneys or ribbons that are related to the intersections of fissures rather than to bedding planes. In all these deposits silver is the most important metal, but there are important amounts of lead and copper in the deposits at Bullion. Gold is usually present, but always in subordinate quantity. As a rule the ore is highly siliceous. The gangue minerals are quartz, barite, and calcite.

The principal sulphides are pyrite, galena, zinc blende, argentite, and chalcopyrite. Stibnite, stromeyerite, gray copper, polybasite, stephanite, and other minerals containing arsenic or antimony are present in some of the ore. Contact-metamorphic silicates are wanting. The ore bodies of this group are without exception near the deep-seated intrusives and are believed to have been deposited by solutions which were given off from these rocks as they cooled.

REPLACEMENT VEINS AND SHEETED ZONES IN LIMESTONE AND IN SHALE.

The replacement veins and sheeted zones in limestone and in shale are in many respects similar to the deposits just described, but instead of chimneys and irregular masses they are thin tabular bodies. The silver deposits at Columbia and Aura, in the Centennial Range, and some of the deposits at Lewis, in the Shoshone Range, belong to this class. The deposits occur in areas of calcareous sedimentary rocks intruded by granodiorites and related igneous rocks. Some of the veins are parallel to the bedding of the country rock, but most of them cut across it. They are clearly related to fissures and zones of movement and are apt to be wider in limestone than in shale, because the limestone is more favorable for openings and more readily replaced by the vein-forming solutions. The sheeted zones are in most respects similar to the replacement veins, but were deposited in two or more narrow, closely spaced, approximately parallel openings, and they usually include more or less of the brecciated country rock. As a rule the ore is highly siliceous. The gangue minerals are quartz, barite, and calcite; the sulphides are pyrite, galena, zinc blende, chalcopyrite, stibnite, and gray copper. Ruby silver and argentite are present in some of the deposits. At Mountain City and Mill Canyon there are some undeveloped prospects of ferruginous oxidized gold ore which probably belong to this group.

FISSURE VEINS IN QUARTZITE.

The fissure veins in quartzite are of considerable economic importance, for they include the gold deposits at Edgemont and at Bull Run and some of those at Dean. These deposits occur in areas of intrusive granodiorite or granodiorite porphyry, but those at Edgemont are more than a mile from known outcrops of igneous rocks. The ore is simple in composition, the gangue is quartz, and the sulphides are present only in small amounts. The primary sulphides are pyrite, galena, and arsenopyrite, with a small amount of chalcopyrite. Silver is present in subordinate quantity, but there is not enough copper to interfere with cyanide extraction. The gold is in the quartz and sulphides and a large proportion of it is free milling.

FISSURE VEINS IN THE OLDER INTRUSIVE ROCKS.

The group of fissure veins in the older intrusive rocks includes the principal deposits at Mountain City, Dean, Grey Eagle, and Mill Canyon. All of these deposits are in clear-cut fissures in granular rocks or in their porphyries, and were formed presumably soon after the intrusives had solidified. The veins were deposited by hot waters which probably rose from the deeper portions of the cooling intrusive mass. Along all these deposits the wall rock shows the effects of hydrothermal metamorphism. Pyrite, sericite, and quartz are everywhere developed in the wall rock near the veins and most extensively near the larger deposits. Calcite is usually associated with the sericite and was presumably formed at the same time. Adularia is not present. In the deposits at Mountain City, Grey Eagle, and Mill Canyon silver is the principal metal, but the ore carries also gold and lead. The sulphide ore is composed of quartz, pyrite, galena, zinc blende, gray copper, argentite, gold, arsenopyrite, and a little chalcopyrite. The deposits in granite at Mill Canyon are of the same general composition as those at Mountain City. At Dean the gold veins in granodiorite carry a very small amount of gray copper.

LATER DEPOSITS (MIOCENE).

After the early igneous activity had subsided there was a period of relative quiet, during which the country was eroded and supplied sediments for the lake lying just east of this area. In this lake the Eocene beds were deposited. In Miocene time volcanism on a grand scale was repeated. The first eruptions were rhyolites and later these were intruded by andesites and at some places covered with basalts. So far as the ore deposits are concerned the rhyolite and andesite only need be considered. Both of these form the country rock for the deposits of this group, but the andesites seem in all places to be the agents of mineralization. All the ore bodies of this group are in the andesites or in rhyolite near intrusive andesite. As the andesites were intruded into rhyolites, which are of Miocene age, and preceded basalts, which are Pliocene, they must have been intruded at or near the close of the Miocene. The ore deposits are therefore believed to be of late Miocene age. Conservatively estimated, they have yielded about $28,000,000. A list of the minerals of these deposits is given below:

Adularia.	Chalcanthite.	Kaolin.	Pyromorphite.
Argentite.	Chalcocite.	Limonite.	Quartz.
Arsenopyrite.	Chalcopyrite.	Malachite.	Sericite.
Azurite.	Chlorite.	Manganite.	Sphalerite.
Barite.	Chrysocolla.	Muscovite.	Stephanite.
Bornite.	Enargite.	Orthoclase.	Stibnite.
Calcite.	Freibergite.	Pyrargyrite.	Tetrahedrite.
Cerusite.	Galena.	Proustite.	Turquoise.
Cerargyrite.	Gold.	Pyrite.	Zinc blende.

The principal deposits of this age may be divided into the following groups: (1) Fissure veins and sheeted zones in andesite; (2) fissure veins and fracture zones in rhyolite.

FISSURE VEINS AND SHEETED ZONES IN ANDESITE.

The deposits in andesite include the silver veins of Tuscarora and the principal deposits at Cornucopia, Burner, Falcon, and Stafford. They are banded siliceous veins which carry a variable amount of sulphides. Some pyrite, galena, and zinc blende are present, and in the richer deposits gray copper, ruby silver, stephanite, and other antimony and arsenic sulphides occur in appreciable quantities. Silver is the principal metal won, but the ore usually contains an important amount of gold.

Hydrothermal metamorphism of the propylitic type is everywhere pronounced in the andesite near these deposits. It is more conspicuous because it produces striking color changes in the rock. Sericite and pyrite are very extensively developed, and at Tuscarora adularia forms in considerable quantities. Between the most-altered phases of the andesite and the fresh, unaltered rock, chlorite, resulting from the decomposition of biotite and hornblende, forms in great abundance and some carbonates are deposited. At such places the rock is dark green and grades into the brownish-gray andesite on one side and into white decomposed sericitic andesite toward the ore deposits. The hydrothermal metamorphism of the andesite is confined to the areas of the ore deposits and was accomplished by the same solutions that deposited the ore in the fissures. The amount of change is directly proportional to the mineralization and the extent of the area affected depends on the distribution of the fissures. The bulk of the deposition was made in the open spaces, but where the walls were strongly fractured they were replaced by the ore minerals. Compared with the alteration of the walls caused by the vein-forming solutions along the veins in granodiorite, the changes along the younger veins in andesite are more extensive and the area affected is greater. There has been less erosion since the ores in andesite were deposited and consequently the ore bodies exposed were formed under thinner cover than the deposits associated with the older intrusive rocks. Lindgren has explained the extensive alteration which is almost invariably shown near the deposits in the younger eruptive rocks by the fact that the light covering at the time of deposition favored porosity and open spaces, which permitted the solutions to penetrate the rocks with greater freedom.

FISSURE VEINS AND FRACTURE ZONES IN RHYOLITE.

The ore bodies in rhyolite occur near intrusive andesite. The group includes the deposits of the Dexter mine, at Tuscarora, the lodes at Gold Circle, and some prospects which have lately been discovered in the Lynn district. Gold is the principal metal, but a small amount of silver is present. Hydrothermal metamorphism is noticeable, but as a rule it is less intense and much less conspicuous than in the andesite. It consists in the main in the development of sericite, pyrite, and secondary quartz, and at the Dexter mine considerable adularia has been deposited. Chlorite is much less abundant than in the altered andesite and carbonates are wanting. The ore is very simple in composition. Most of it consists of quartz and pyrite. The deposition has usually taken place in many small openings in a broad zone of fracturing rather than in wide open spaces.

PLACER DEPOSITS.

Placer deposits are relatively unimportant in this area and have not been discovered at most of the camps. They have been found at Tuscarora, Aura, Van Duzer Creek, and Lynn. At Van Duzer Creek and Aura they are derived from the disintegration of the lode deposits of the earlier group; at Tuscarora and Lynn from the younger lodes in the Tertiary eruptive rocks.

The gold placers at Tuscarora were worked extensively in the seventies and are said to have yielded about $7,000,000. At present no work is being done except that carried on by a few Chinese. The gold occurs as dust and as nuggets of considerable size. Its source is presumably some gold lodes which occur in rhyolite and andesite to the north and west of the diggings. A large acreage of ground west of Tuscarora has been located and sampled with drills. It is said that much of this ground will pay to work with dredges.

At Van Duzer Creek, which is between Aura and Mountain City, placer mining was carried on during the summer months for a number of years. Two small reservoirs have been built in this creek and steel pipes have been laid from these to supply water for monitors. There are few large bowlders in this gulch, and when the water supply is sufficient operations may be carried on with success. The source of the gold consists of some undeveloped veins at the head of the stream. The mines were not worked in 1908.

In the Lynn district some of the gulches which head in an area of mineralized rhyolite carry placer gold and a few hundred dollars was recovered in pans and rockers during the season of 1908.

An attempt has been made to work the gravels of Bull Run Basin, near Aura, and an extensive system of ditches and flumes has been installed. After a few hundred yards had been washed out the project was abandoned. It is said that some gold was recovered, but on account of the size of the bowlders in the gravels their exploitation was unprofitable. Some notes on the placers are given where the camps are described.

PROSPECTING.

From what has been said it is clear that the deposits of the older group are confined to the granular rocks and deep-seated porphyries and to the sedimentary formations in areas which have been intruded by those rocks. It would be worth while to scrutinize closely the country along the margins of the granular intrusive rocks and especially the limestones and quartzites some distance away. These rocks may be but little altered in the immediate vicinity of the deposits, and except where garnet zones are developed there may be in the country rock itself no conspicuous changes indicating mineralization. Silicified outcrops, iron-stained rocks, or gossans may mark the presence of the older deposits in sedimentary rocks. The ore-bearing solutions seem to have been capable of traveling farther with their burden when they were moving in fissures formed in quartzite than in the limestone. Quartzite is not readily replaced, especially by solutions rich in silica, and consequently the deposition does not occur until the solutions are cold enough to deposit by simple precipitation rather than by interchange of molecules with the calcareous wall rock. Deposits in quartzite are therefore likely to be farther from the intrusive granular rocks than the deposits in limestone. At Edgemont the lodes fill open spaces in quartzite and are more than a mile away from the nearest known intrusives. Along the lodes in the granodiorites the country rock is usually altered and leached white or pale green, but such alteration may not extend far from the lodes.

As already stated, a large part of the area is covered by late Tertiary lava flows, chiefly glassy rhyolites and vesicular basalts. The basalts are not known to contain any ore deposits, and the rhyolites are probably barren except where cut by later intrusives. The surface flows do not seem to have been agents of mineralization. If the magmas which formed them carried ore-bearing solutions these must have escaped during eruption and were lost. The deposits associated with the eruptive rocks of the younger period are in the intrusive andesites or in rhyolites cut by such intrusives. In prospecting for these deposits search should be made for the andesite, a dark-gray or brown rock easily recognized in the rhyolite areas. It is not so dark as the basalt and does not contain the green olivine crystals which are characteristic of that rock.

The later Tertiary deposits are limited also to areas marked by strong hot-water action. The prospector who is familiar with the deposits in the later eruptive rocks is well aware that they are confined to the leached areas. The "kindly look of the rock," or the dull white appearance resulting from devitrification, sericitization, and kindred processes, is quickly recognized. At many places these light-colored chalky areas may be distinguished far away, especially where they are located on bare slopes and are surrounded by the darker andesite or by the glassy pinkish rhyolites or dark obsidians. Such areas of mineralization may yet remain to be discovered, but those which are now unknown are probably in obscure places partly covered by vegetation, soil, or rocky débris.

MOUNTAIN RANGES AND MINING DISTRICTS.

OWYHEE BLUFFS.

GENERAL FEATURES.

The Owyhee Bluffs, which are in the northwestern part of the area covered by this reconnaissance, form a lofty ridge trending northeastward toward the Independence Mountains. Rose Mountain, one of the highest summits of this ridge, reaches an elevation of 7,949 feet, or about 2,000 feet above the level of Squaw Valley, which lies to the southeast of the bluffs. This valley is a broad expanse of agricultural land almost completely shut in by hills and low mountain ranges. Northward from the Owyhee Bluffs the country slopes very gently to the Owyhee Desert, a great expanse of level country which extends northward far into Idaho. Through this great plain the small tributaries to Little Humboldt River and to the forks of Owyhee River have sunk their channels, and here and there small rounded hills relieve the monotony of the otherwise featureless landscape. Owyhee Desert is better watered than the name implies and during a part of the year affords subsistence for live stock.

GEOLOGIC FEATURES.

The Owyhee Bluffs[a] are made up almost exclusively of eruptive rocks of Tertiary age. The bedded rhyolites have by far the greatest distribution, forming the larger part of the crest of the ridge and extending downward to the base of the slopes. The rhyolites belong to the extensive series of flows which cover the larger part of the area of this reconnaissance north of Humboldt River. About 5 miles northeast of Midas, along the floor of a small canyon, there is a small area of shaly limestones surrounded by rhyolite, and so it is inferred that these flows rest upon the eroded surface of sedimentary rocks. The rhyolites are cut by andesite dikes and here and there are overlain by andesite flows, one of which is capped by rhyolite.

Squaw Valley borders the Owyhee Bluffs on the southeast and, like many of the smaller areas of low land between the mountain ranges, is covered for the most part with a deposit of Quaternary gravels, but in the lower end of this valley there are some stratified deposits of volcanic ash[a] which on lithologic grounds have been referred to the Humboldt formation, of Pliocene age.

---

a Emmons, S. F., U. S. Geol. Expl. 40th Par., vol. 3, 1870, p. 612.

LOCATION AND HISTORY.

The Gold Circle district is situated in the hilly country along the southeastern slope of the Owyhee Bluffs, near the edge of Squaw Valley. It is about 45 miles north of Battle Mountain and approximately the same distance from Golconda, and is connected with both of these stations by stages which make round trips three times a week. In the summer of 1907 gold was discovered on several claims, and in March, 1908, as a result of a number of rich strikes, the district experienced one of the rushes which is characteristic of the method of settlement of mining camps in Nevada. A town site was laid out at Midas, and within a few weeks some 1,500 persons were established in this town. After the first excitement had passed away a majority of the newcomers left, and in September, 1908, the population of the camp had decreased to about 250 persons. Several of the claims were under development, and a number of lodes were being prospected with more or less success. A few tons of rich ore have been shipped to smelters, but the bulk of the ore that has been developed is not of a grade to pay the shipment charges, which are necessarily high, as they include a long wagon haul. The deepest shaft is sunk 200 feet; several other shafts are down 100 feet; and three or four tunnels have been driven to depths approximately 100 feet below the surface. Ground was broken in September, 1908, for a 10-stamp custom mill, which it was planned to erect at once, and two mining companies were contemplating the erection of mills in the near future.

GEOLOGY.

General outline.—The rocks of the Gold Circle district are rhyolite flows and flow breccias which are cut by dikes of andesite and overlain here and there by andesite flows. The rhyolites, which are the oldest rocks exposed in the district, cover the greater portion of the area. They occur in considerable variety, but the most common is a light-colored, dense, streaked rock composed in the main of a glassy or devitrified groundmass which contains scattered phenocrysts of feldspar and quartz. Other phases of the rhyolite are perlitic and some are vesicular. In the vicinity of Queen Canyon, east of the Esmeralda claims, the vesicles of rhyolite are filled with amygdules of beautiful amethystine quartz.

In Queen Canyon and at several other places in the Gold Circle district the rhyolite is highly fissile and thinly bedded, presenting the appearance of a silicified shale. The shaly appearance is probably due to banding that was developed as the rhyolite flowed and is emphasized by subsequent weathering and the deposition of iron oxide along the parting planes. The rhyolites include also flow breccias which contain many angular fragments of streaked rhyolite cemented by a matrix of glassy rhyolite. This brecciation is due to movement of the flow while it was still more or less fluid, but after a crust had formed on the surface. The brecciated rhyolite, after weathering or alteration by hot waters, may closely resemble a friction breccia along a vein, and it may easily be mistaken for ore, unless the fact that the cementing material is glass and not quartz is noted.

The andesite which outcrops at many places in the district is a dark, fine-grained porphyritic rock, cutting through the rhyolite or forming flows interbedded with it. Under the microscope it is seen to be composed of a brown glassy groundmass containing crystals of andesine and labradorite feldspar, a considerable amount of augite, some magnetite, and a little quartz. Augite and the brown ground-mass are partly altered to serpentine, calcite, and sericite. The feld-spars contain too much soda for basalts, and no olivine was found in any of the thin sections. Dikes of andesite in rhyolite are well exposed on the Dixie claim, about half a mile northwest of Midas. The andesite caps the rhyolite, at some places forming only a thin veneer above it, as is shown on the Iron Mask claim, 1 mile east of Midas, and on the hillside to the south of this claim. Some of the andesite is highly vesicular, and the larger portion of it undoubtedly forms sur-face flows, the dikes representing the vents through which the flows rose to the surface. About 1½ miles N. 15° E. of Midas, a few rods north of the Elko Prince Annex claim, there is a hill which is com-posed almost entirely of andesite. The rock forming the lower por-tion of the hill is solid porphyritic andesite; the upper 50 feet is highly vesicular and probably represents the upper portion of the same flow. Above the vesicular portion of the andesite is a bed of rhyolite, which represents a flow that was poured out subsequent to the extravasation of the andesite. In the country to the north of the Gold Circle district another bed of rhyolite was noted above andesite. Although the main mass of the rhyolite was erupted before the andesite, it is very clear that some of it is later than the andesite. This sequence is suggested also by a dike of felsitic quartz porphyry similar to rhyolite in composition which cuts andesite about one-fourth mile northwest of the Midas mine.

Hydrothermal metamorphism.—In a view of the rugged south-eastern slope of the Owyhee Bluffs from the south, the Gold Circle district, including an area some 3 miles square, stands out in sharp contrast with the surrounding country. The rocks in this area are leached to a chalky white, stained to a light brown here and there by iron oxides. In the fresh glassy rhyolites which surround this area shades of pink and greenish gray predominate, the coloring matter being due to a very small amount of iron present in the glass. In the region of the ore deposits, which is in a broad way coincident with the leached area, hot solutions have soaked into the country rock, causing devitrification of the glass and other mineralogical changes. Near the lodes, where the action was more intense, pyrite, quartz, chlorite, and sericite have been formed in the rhyolite. The few slides studied indicate that these minerals are rather closely restricted to the country rock within a few feet of the lodes and that devitrification has taken place farther away. The changes in andesite are less intense than in rhyolite; some feldspars are slightly sericitized and augite is partly altered to calcite and chlorite. Calcite, which is present in considerable quantity in the altered andesite, seems not to have formed in rhyolite, and that in the andesite may have been formed altogether by surface waters subsequent to the deposition of the ores.

Fissuring and faulting.—There has been considerable fissuring and faulting since the eruption of the rhyolite and andesite. All the ore deposits are related to planes of movement. At the Rex mine, on the Gold Circle claim, in the Sleeping Beauty tunnel and elsewhere, andesite and rhyolite are in faulted contact. The geologic sketch map (fig. 3) indicates the approximate distribution of the rhyolite and andesite. In work of a more detailed character the separate rhyolite flows could probably be distinguished and the details of the faulted structure could be worked out, but this was not done in the few days which were given to the study of the district. The lodes are plotted with a greater degree of accuracy than the boundaries of the geologic formations, which, at many places, were not traversed.

ORE DEPOSITS.

General features.—The deposits in the main are replacement veins and sheeted zones in rhyolite, which are located along prominent slickensided planes of movement. All the fissures strike northwestward and are with a few exceptions approximately parallel. In general they dip from 65° to 85° NE. In the commonest type a few inches of high-grade iron-stained siliceous ore occurs here and there along the slip planes, and in places the surrounding country rock for a distance of several feet is shattered and seamed with veinlets of quartz carrying gold. Thin drusy cavities with well-formed crystals of colorless quartz are found locally in these veinlets, and at some places this quartz is banded with a black silver-rich mineral, probably argentite. The rhyolite near the vein is devitrified, silicified, stained with iron oxide, and at many localities replaced by ore. In the St. Paul mine, where dark silver-bearing sulphides and pyrite occur in banded ribbons alternating with quartz and parallel to the walls of the vein, the deposit is a simple fissure filling, but in most of the lodes where the sulphides are shown the original openings were small and the deposition was mainly through replacement or impregnation

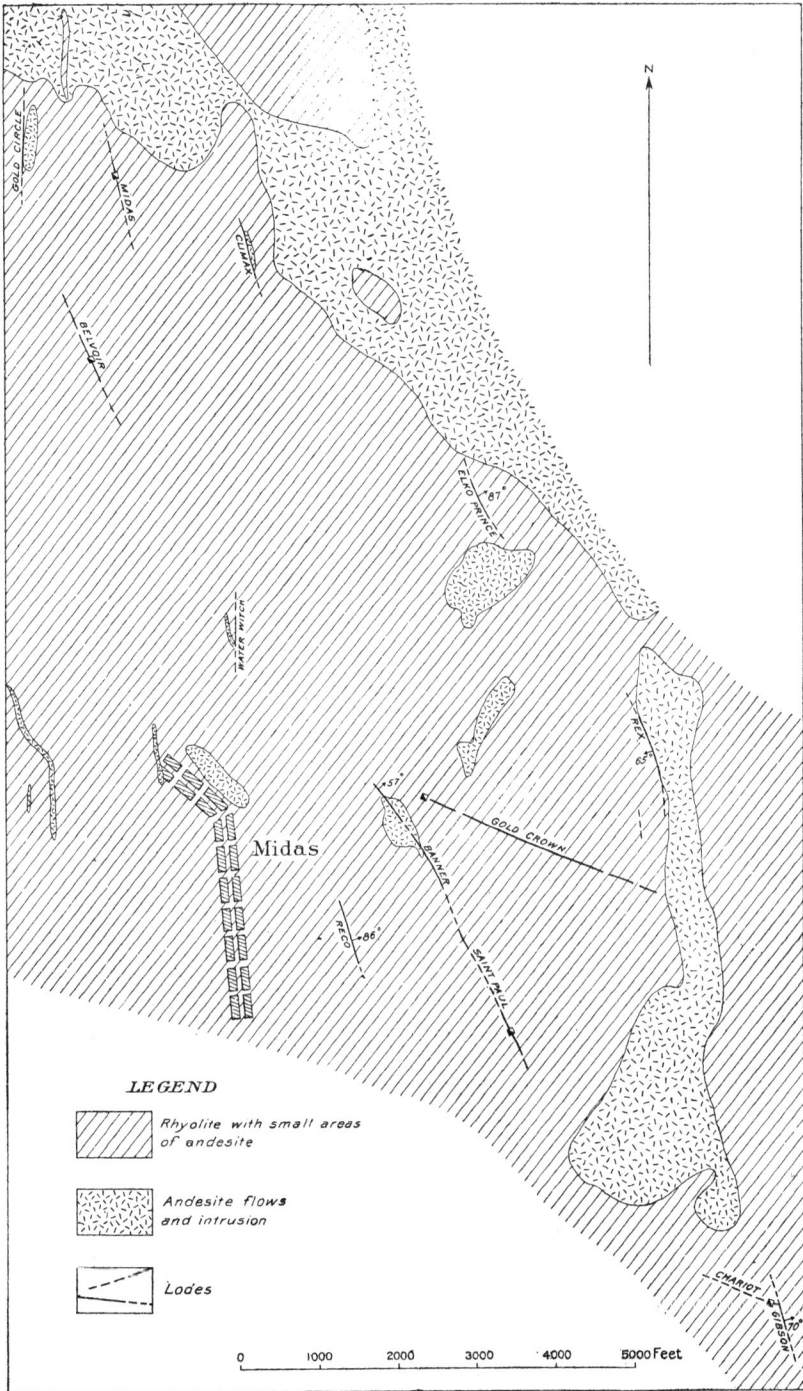

FIGURE 3.—Geologic sketch map of Gold Circle mining district.

of the rhyolite. The primary ore minerals are pyrite, quartz, gold, and probably argentite. The secondary minerals are quartz, iron oxide, manganese oxide, and horn silver. The gold is almost without exception associated with pyrite, with iron oxide, or with quartz highly stained with iron or manganese. At some places the sulphides begin to appear within a foot or two of the surface. The depth of the partly oxidized zone is from 100 to 150 feet below the surface.

Prospecting the lodes.—Well-defined fissures are very conspicuous in many of the replacement deposits. Some of these were formed before the ore was deposited, as is shown by the thin tabular bodies of oxidized ore which occur along some of the fissure planes and by the surfaces of the planes, which are corroded so that they do not show the polished striated surfaces that movement planes commonly exhibit. Along some of the lodes there are, however, slickensided planes which are clearly later than the ore. These are usually polished smooth and striated, and the ore along them is more or less brecciated. In some of the lodes this ore is ground to a gouge containing well-rounded fragments of quartz.

In developing a lode it is important to follow the fissures which were formed before the ore was deposited rather than those which are of later origin. In the oxidized zone it is difficult to distinguish the two classes of fissures, especially where there has been movement before and after deposition approximately along the same plane. The corrosion of the surface of the fissure and the presence of a thin tabular body of uncrushed quartz may distinguish these fissures from the slickensided planes of postmineral movement, which, in places, may carry crushed ore. Probably most of the faulting that brings the andesite into steep contact with the rhyolite took place after the deposition of the lodes, for, as a rule, such fault contacts show slickensiding or striæ, in contrast with the etched or roughened surfaces of fissures along which the uncrushed ore occurs. Where the lodes are walled on one side by the andesite they are, as a rule, highly crushed, but the andesite is relatively fresh, or at least is not nearly so much altered as the rhyolite. A faulted contact between these two rocks should be investigated with a view to finding a vein, for many of the planes of later faulting followed the zone of earlier fissuring, but it is not good prospecting to drift for great distances along faults which do not show mineralization or crushed ore, or which do not pan gold in the soil along the surface. Developments thus far have not shown any faults which cut across the lodes and displace the ore.

Ore shoots.—Notwithstanding the continued activity of lessees since the early discoveries, the search for shipping ore has thus far proved disappointing. Rich pannings and small bunches of high-grade ore are found near the surface in many places, but the present state of the mines indicates that the production will have to come in the main from lower-grade deposits of milling ore. Some of the fissures along which the ore is found are regular and persistent, and have been opened here and there for half a mile or more along the strike, but these, where developed, do not carry ore of milling grade throughout their length, and development is not sufficient to show how extensive the ore shoots are. In the Rex mine a body of good milling ore 180 feet long has been developed on three sides to a depth of 65 feet, and there is also a considerable tonnage, partly in the sulphides, in the workings along the Gold Crown lode. A number of leases are in ore of milling grade, but these ore bodies have not been sufficiently developed to be regarded as ore in sight. On several of the undeveloped claims the surface showings seem sufficient to warrant further prospecting for milling ore.

Secondary enrichment.—The primary ore is auriferous pyrite and quartz, with which are associated a silver-bearing sulphide and other minerals. As the surface is worn away such ore is oxidized and the sulphur, together with most of the silica and iron and some of the gold, is carried away, but a larger proportion of the gold remains. As a result of this process there is likely to be a concentration of the gold in the upper part of the deposits, but to what extent such concentration has taken place in this district is not known. Some of the little seams of rich ore are solid and appear to have suffered slight change except oxidation. Some very good values have been found in the sulphide ore about 200 feet below the surface of the Gold Crown shaft, but there has been so little exploration in the primary sulphide ores that it is not possible to compare its value with that of the oxide ore.

RÉSUMÉ OF GEOLOGY.

A study of the Gold Circle district shows the following geologic history. In Tertiary time, probably in the Miocene, extensive flows of rhyolite were poured out upon a surface of Paleozoic sedimentary rocks. Subsequently the rhyolites were fissured and through these fissures andesite flows rose to the surface and covered the rhyolite. A portion of the magma remained in the fissures, forming dikes. The andesite was in turn cut by fissures which were filled with an acidic magma, of which one portion flowed out upon the surface and formed rhyolite and another portion, solidifying in the fissures, formed dikes of quartz porphyry. After the eruption of the andesite the country rock was strongly fissured, most of the planes of movement striking northwestward and dipping steeply to the northeast. Along some of these fissures auriferous pyrite and quartz with silver-bearing sulphides were deposited, the solutions dissolving portions of the country rock and replacing it where conditions were favorable with ore and other minerals. From the fissures the solutions spread out into the country rock, causing devitrification and other changes. A second fissuring with some displacement occurred after the ores were deposited. These movements were mainly along the lodes and brecciated the quartz and sulphides. As a result, the ore shoots, which were already irregularly spaced along the earlier fissures, were strung out along a plane of later movement. As the rocks were eroded the ores were oxidized, pyrite changing to limonite, sericite and feldspar to kaolin. Hydrous silica was deposited in crevices, where it is associated with free gold and manganese oxides.

MINE DESCRIPTIONS.

Rex mine.—The Rex mine is on the eastern slope of a low ridge about a mile east of Midas. A shaft driven at an inclination of 64° is sunk to a depth of 65 feet. From the bottom of the shaft a drift is run 30 feet to the north and 120 feet to the south, with short crosscuts here and there. The lode is along a fault between rhyolite and andesite. It strikes about N. 15° W. and dips 66° W. It is a zone of crushed, silicified, iron-stained rhyolite from 5 to 16 feet wide and carries, according to C. G. Rothschild, from $5 to $28 a ton. The rhyolite is a dense, light-colored rock with a few phenocrysts of feldspar and quartz. In the lode it is silicified, iron stained, and cut by veinlets of quartz. The ore developed is highly oxidized, but a little pyrite is present in the bottom of the mine. There is a well-defined slickensided plane of movement along the foot wall between the andesite and the ore, and other fissures approximately parallel to this one cross the ore zone. The ore is restricted to the altered rhyolite, and the andesite, even where greatly crushed, is said to be barren. For this reason it seems probable that the displacement which brought the rhyolite and andesite into contact occurred after the deposition of the ore. Figure 4 is a plan of the Rex mine on the 65-foot level.

Gold Crown lode.—The Gold Crown lode between the Rex mine and Midas has been developed in a number of shafts, pits, and tunnels for a distance of nearly 3,000 feet along the strike. It is a zone of shattered rhyolite which strikes N. 67° W. and dips about 65° N. Wherever the lode has been developed there is a well-defined fissure which at some places is slickensided and carries gold values in crushed quartz. At the Snowstorm lease, at the west end of the lode, half a mile east of Midas, a vertical shaft is 84 feet deep and short levels are turned 70 feet and 84 feet below the surface. The shattered rhyolite for a width of 25 feet is said to carry milling ore. In the Climo lease, farther east on the lode, several pits and short tunnels expose a regular fissure along which some gold values have been obtained. Still farther east the lode is developed in the lower tunnel of the Gold Circle Crown Mining Company, where it is a wide zone of shattered rhyolite, through which considerable pyrite is disseminated. East of this tunnel, along the strike of the lode and higher on the hill, is a second tunnel driven by the same company, and in this the lode is most extensively developed. The upper tunnel is a crosscut for a distance of 55 feet to a point where it encounters along the foot wall of the lode a smooth fissure that strikes S. 60° E. and dips 65° NE. This fissure is followed in the tunnel for a distance of 400 feet, and a shaft 200 feet deep is sunk in the hanging wall of the same fissure and intersects the tunnel at a depth of 80 feet. Small bodies of rich ore, from 1 to 10 inches wide, are found here and there along the fissure and the hanging-wall rhyolite is shattered, crushed, and cemented with veinlets of quartz and pyrite. A zone of the crushed rhyolite about 3 feet wide is said to be good milling ore. Some crushing has taken place since the ore was deposited, for fragments of quartz rounded by attrition are found here and there along the fissure in a mass of crushed leached rhyolite. The oxidized ore extends below the surface to a depth of about 110 feet, where the sulphides are encountered. The values in the oxidized and in the sulphide ore, so

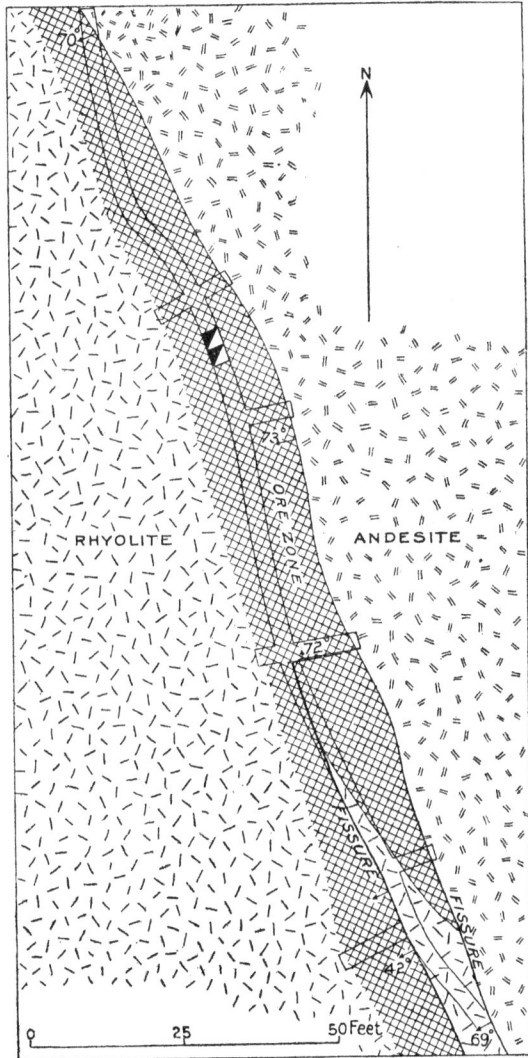

FIGURE 4.—Plan of Rex mine, 65-foot level, Midas (Gold Circle) district.

far as developed, are said to be approximately the same. The Emancipation lease is about 1,500 feet east of the Crown shaft and on the strike of the Gold Crown lode. Ninety feet below the collar of an inclined shaft driven in rhyolite there is a fissure which strikes southeastward and dips 58° NE. and carries pockets of good milling ore.

St. Paul-Banner lode.—The St. Paul mine is about a mile southeast of Midas. A shaft inclined 76° E. is sunk on the lode to a depth of 132 feet and levels are turned 60 and 120 feet below the surface. The country rock is rhyolite and rhyolite flow breccia. The rhyolite near the lode is altered to kaolin and sericite and contains small crystals of pyrite, but these are said to be barren. The lode is a simple fissure filling which has a maximum width of about 12 inches and carries high values in gold and silver. It strikes northwestward and dips steeply to the northeast. The ore is oxidized to a depth of 70 feet. Below this depth it consists of banded quartz and dark argentiferous sulphides with faint red bands that are probably ruby silver. This deposit at the St. Paul mine differs from the other ore bodies of the Gold Circle district, the ore being a banded fissure filling rather than a replacement vein. A few tons of ore carrying $100 to the ton have been shipped to smelters.

To the northwest, in the line of the strike of the St. Paul lode, exposures show values in several open cuts toward the Banner ground, and in one of these a little horn silver was found. On the Banner claim a fissure which is presumably the same as that of the St. Paul is exposed in three shafts. In the north shaft fine-grained andesite occurs on the west side of the fissure, which just north of this point passes through a flow of andesite that caps the rhyolite. On the Ripsaw claim, still farther north and in the strike of the Banner lode, a fissure with approximately the same strike is exposed in a long trench, where it dips 60° E. The Reco lode (No. 5) is 1,300 feet southwest of the Banner lode and approximately parallel to it. A shaft is sunk on a fissure in rhyolite which dips 86° E.

Golden Chariot claim.—On the Golden Chariot claim, 1 mile southeast of the St. Paul mine, three pits are sunk on the Gibson lode, which strikes a few degrees west of north and dips steeply eastward. Along the lode the rhyolite is highly shattered and stained with iron oxide. At the bottom of one of the pits there is a streak of rich ore composed of soft, rotten rhyolite, black manganese oxide, and hydrous silica, with numerous small flakes of free gold. On the Chariot vein near by a 50-foot shaft is sunk in shattered decomposed rhyolite, which carries pockets of rich ore.

Esmeralda mine.—The Esmeralda claim is about 3,500 feet south of the Golden Chariot. On the Charters-O'Byrne lease on this claim a 45-foot shaft is sunk on the promising outcrop of a lode which dips 76° SW. The strongly shattered rhyolite for a width of 3 feet along the lode carries low-grade gold ore. On the Riddle lease on the same lode, about 300 feet to the southeast, a shaft is sunk 50 feet in shattered iron-stained rhyolite, which is strongly mineralized and carries pockets of high-grade gold ore.

Water Witch mine.—The Water Witch claims are about half a mile north of Midas. In the Benan lease, on the Water Witch fraction, a shaft is sunk in rhyolite on a sheeted zone which strikes a few degrees west of north. Along a prominent plane of movement there is from 1 to 12 inches of rich gold ore. A 3-foot zone of shattered rhyolite is said to carry $20 to the ton. The screenings of the dump are said to carry shipping values.

Elko Prince lode.—The Elko Prince lode, about 1 mile northeast of Midas, is in altered rhyolite, strikes northwestward, and dips steeply to the northeast. It is developed in a shaft about 100 feet deep and in several surface pits near by. In places on the surface the lode is a sheeted zone of iron-stained rhyolite, but in the shaft it is a banded siliceous filling of an open space about a foot wide. The ore is said to carry $20 a ton in gold and silver.

Midas mine.—The Midas claim is about $1\frac{1}{2}$ miles north of the town of Midas. A shaft 100 feet deep is sunk in rhyolite on a sheeted zone which is from 2 to 5 feet wide and is said to carry $8 to the ton in gold. The rusty ore from a small pay streak shows numerous specks of free gold.

Gold Circle claim.—The Gold Circle claim is about 1,200 feet northwest of the Midas. A zone of crushed silicified and highly iron-stained rhyolite along a fault between andesite and rhyolite pans free gold liberally. This deposit was discovered only a few days before the camp was visited by the writer and not more than 10 feet of work had been done.

Belvoir claim.—On the Belvoir claim, one-half mile south of the Midas, a sheeted zone of rhyolite is exposed in small pits and trenches. Some of the small fissures are filled with crushed rhyolite, cemented with iron-stained quartz which carries gold and silver.

Eastern Star mine.—The Eastern Star mine is on Frazier Creek about $4\frac{1}{2}$ miles east of Midas, in an area of white devitrified rhyolite which is similar in appearance to that of the Gold Circle district. The two bodies of leached rhyolite are not directly connected, however, for a large area of fresh vitreous rhyolite lies between. A tunnel is driven northward for 160 feet along the strike of a zone of silicified iron-stained rhyolite which outcrops boldly on the summit of a low ridge. It is cut by three or four parallel veinlets up to 10 inches wide, which carry ribbons of quartz and argentite and in places show a liberal amount of free gold. Between the veinlets and for some distance on either side the rhyolite is impregnated with finely divided pyrite and other opaque minerals, so that a considerable mass of it is as dark as andesite.

INDEPENDENCE RANGE.

GENERAL FEATURES.

The Independence Range is a compact group of lofty mountains east of the Owyhee Bluffs and separated from them by a low pass which is crossed by the road from Gold Creek to Burner. The range may be regarded as a northward extension of the Cortez Range, but the two are separated by a relatively low saddle between the headwaters of Soldier Creek and Independence Valley. Several of the highest mountains reach elevations from 9,000 to 10,000 feet above the sea, or about 4,000 feet above the level of Independence Valley.

As shown in the atlas of the Fortieth Parallel Survey, the range is made up mainly of thick rhyolite flows which here and there are eroded away to expose large outcrops of sedimentary rocks, chiefly the Weber quartzite. The rhyolite extends westward a great distance from Tuscarora and covers a large area surrounding Squaw Valley, including Gold Circle and other camps near by. At Tuscarora, Cornucopia, Good Hope, and Falcon the rhyolite is cut by large bodies of intrusive andesite. A body of granodiorite occurs near the headwaters of Willow Creek.

The mining camps situated in the Independence Range are Tuscarora, Cornucopia, Good Hope, and Falcon. The ore deposits are silver-bearing fissure veins of the Tertiary group and are in andesite or in rhyolite near intruding andesite. No deposits have yet been discovered in the sedimentary rocks or associated with the granodiorite intrusive rock.

TUSCARORA.

HISTORY.

Tuscarora is situated on the southeastern slope of Mount Blitzen, at an elevation of about 6,200 feet above sea level. It is near the west margin of Independence Valley, a broad area of flat hay land drained by Owyhee River, and is about 50 miles northwest of Elko, with which it is connected by a daily stage that also connects at Tuscarora with stages for Edgemont, Aura, Mountain City, and other northern points.

Placer deposits were found at Tuscarora in 1867, and several years later rich silver veins were discovered. In the seventies and eighties a number of silver mines were opened and a large production was sustained for a number of years. Most of the ore was milled at Tuscarora; only the very high grade ore was shipped to smelters. The six silver mills which were in operation employed the Reese River process, by which the ore was dried, stamped, roasted with salt, and amalgamated in silver pans. The Grand Prize mill employed a combination process, the ore being concentrated over vanners and the concentrates roasted and subsequently amalgamated in pans with the raw tailings. The silver mills had an aggregate of 80 stamps and are said to have given a satisfactory extraction.

The Dexter mine, which is a large deposit of low-grade gold ore, was discovered after the silver mines had been producing for several years. This mine was worked until 1898, when operations were discontinued on account of the great volume of water which was encountered. The Dexter ore was treated in the Dexter mill, originally a 40-stamp amalgamating mill operated by electric power. Subsequently four Ellis tables and fourteen cyanide tanks with a capacity of about 450 tons were added to treat the tailings. In 1908 a second cyanide plant was installed.

Since 1898 there has been some leasing on the upper levels of the old silver mine, but extensive mining operations have been discontinued. In 1907 Arthur A. Brownlee obtained options on nearly all the mines and organized the Tuscarora Nevada Mines Company. This company spent considerable money sampling the mines and dumps, and is said to have found a large tonnage of low-grade ore in the old workings of the Dexter mine. The company plans to unwater the mine, to build a large cyanide plant, and to undertake many other improvements.

Accurate figures for the production of Tuscarora are not at hand. Various estimates range from $25,000,000 to $40,000,000. Most of this was obtained between 1872 and 1886 and the larger portion is silver. The gold placers are reported to have yielded $7,000,000 and the Dexter mine $5,000,000 in gold, the various silver mines being credited with the remainder.

When the camp was visited in 1908, practically all the mines were inaccessible. At the Dexter, which is situated at the lower edge of the mineralized area, the water table was within 20 feet of the surface and only some shallow workings and open pits could be entered. In the silver mines, which are at slightly higher elevations, a few workings above the 100-foot level were accessible.

The country rock of the Tuscarora mines is rhyolite and andesite porphyry, which at many places are covered by a thin layer of Quaternary gravel. Most of the deposits are in the porphyry, which in the vicinity of the mines is highly altered. It is the propylite of the Fortieth Parallel Survey. High on the southeastern slope of Mount Blitzen relatively fresh andesite outcrops at several places. The freshest andesite is composed of a dense greenish groundmass containing phenocrysts of andesine, orthoclase, hornblende, and biotite. Toward the mineralized area the andesite is greatly altered. At some distance from the veins chlorite is formed in great abundance and the rock is dark green, but in the highly mineralized area and within a few yards of the veins sericite and iron pyrite have been extensively deposited by replacement. On oxidation this rock alters to a brown iron-stained porphyry, which constitutes a large part of the various mine dumps. Some specimens of what seemed to be the least-altered phases of the porphyry in the mineralized area proved, on examination, to be rich in quartz and orthoclase. It will probably be found that the porphyritic rock commonly regarded as andesite consists of several related kinds of rock, the separation of which will require great patience even under favorable conditions.

The rhyolite is when fresh a white or greenish dense or almost glassy rock; at many places it is a flow breccia. In the vicinity of the ore bodies it is extensively altered, and even the freshest specimens show, under the hand lens, a considerable amount of secondary pyrite.

In the open cut at the Dexter mine the relations of the andesite and the rhyolite are distinctly crosscutting, one rock intruding the other in a very irregular manner. Owing to hydrothermal action and subsequent oxidation it is not possible to ascertain which rock intrudes the other, but if the rhyolite is a flow, as is indicated by the banded shaly phases, then the andesite is the later rock and is intruded into the rhyolite.

ORE DEPOSITS.

The ore deposits, so far as could be ascertained from the limited observations which were possible when they were visited, are silver lodes in andesite, fractured zones or stockworks of gold ore in rhyolite, and gold placers.

Silver lodes.—The silver veins occur in an area of highly altered andesite, which contains small masses of rhyolite. Most of them strike a few degrees west of north and dip toward the west. The gangue is mainly quartz. The ore minerals are ruby silver, enargite, and other silver sulpharsenic and sulphantimony minerals, silver glance, galena, pyrite, and arsenopyrite. A little chalcopyrite and bornite were noted, but these are not abundant. Here and there is a little malachite, but copper is present only in small traces—as a rule, much less than 1 per cent. Shoots of gold ore are found in some of the silver lodes and most of the silver ore contains gold. Near the surface there was much horn silver and native silver. A single block of horn silver from the Commonwealth mine is said to have sold for $30,000. The veins are fissure fillings between the walls of porphyry, and in some of them the ore surrounds numerous fragments of brecciated country rock. Locally the wall rock is replaced by workable ore, and at many places in the vicinity of the veins the country rock is said to carry low values in silver and gold.

The Navajo lode, which is about one-fourth mile west of Tuscarora, was the most productive system of veins. It strikes about N. 80° W. in the Navajo ground and it has been followed northwestward for about a mile. The deposits of the Navajo, Belle Isle, North Belle Isle, Nevada Queen, Commonwealth, and North Commonwealth mines are on this lode. At the North Commonwealth it bends and strikes about N. 60° W. The total production of these mines is said to have been about $15,000,000. East of this lode, and situated in the main on lodes which are approximately parallel to it, are the deposits of the Independence, P. & P., Eira, Silver Prize, Buckeye, De Frieze, Grand Prize, and other mines. According to report, most of the larger ore shoots were at the junction of fissures and pitched toward the northwest.

In the North Belle Isle, on the 70-foot level, several narrow fissure veins strike north and dip from 35° to 80° W. These carry values in silver and gold up to several hundred dollars a ton, and where they join, about 20 feet below the 70-foot level, they make a shoot of ore several feet wide from which $1,000,000 is said to have been taken.

Gold deposits.—The most important gold deposit at Tuscarora is that of the Dexter mine. This is situated near the contact of rhyolite and andesite and dips northward at a low angle. A zone, mainly in the rhyolite, is strongly fractured for a distance of 1,400 feet east and west and about 200 feet north and south. Pockets of rich ore occur here and there in this zone, and it is said that the whole mass of this rock could be worked profitably by cyaniding on a large scale. The deposit is crossed by numerous closely spaced veinlets of quartz, which strike in all directions. In many of them the fissures are not completely filled, and the centers of the veinlets contain long, narrow druses lined on both sides with well-formed crystals of quartz half an inch long. Locally there is considerable adularia in crystals up to one-fourth inch long, deposited in drusy veinlets with quartz. These veinlets are said to carry high values in gold. The rhyolite of the fissured zone is strongly impregnated with pyrite, even the freshest specimens showing much pyrite under the hand lens. The gold seems to have been deposited through replacement of the rhyolite and also in numerous small open spaces. The deposit is crossed by several faults which strike northeastward and dip toward the northwest. The throw of all these faults is said to be toward the south; the hanging wall on the west side of the fault has moved downward, causing an offset of the ore zone toward the south. The faulting is therefore normal.

Some gold placers are located from 1 to 3 miles west of Tuscarora. They were extensively worked in the seventies and are said to have produced $7,000,000, but little work has been done of late years except that carried on by a few Chinese. The deposits were worked mainly by ground sluicing. The gold occurs as dust and as nuggets of considerable size. One of these, about one-half gold and one-half quartz, weighed 9 ounces, and many nuggets have been found which weighed more than an ounce. The source of the gold is presumably some gold lodes which occur to the north and west of the diggings.

At the Rose mine, about 1½ miles west of Tuscarora, several low-grade quartz veins have been found, and the country rock, which is altered porphyry, is said to carry appreciable values in gold.

On Beard Hill, 2 miles southwest of Tuscarora, the Surprise group is situated just above some old placer workings. The country rock

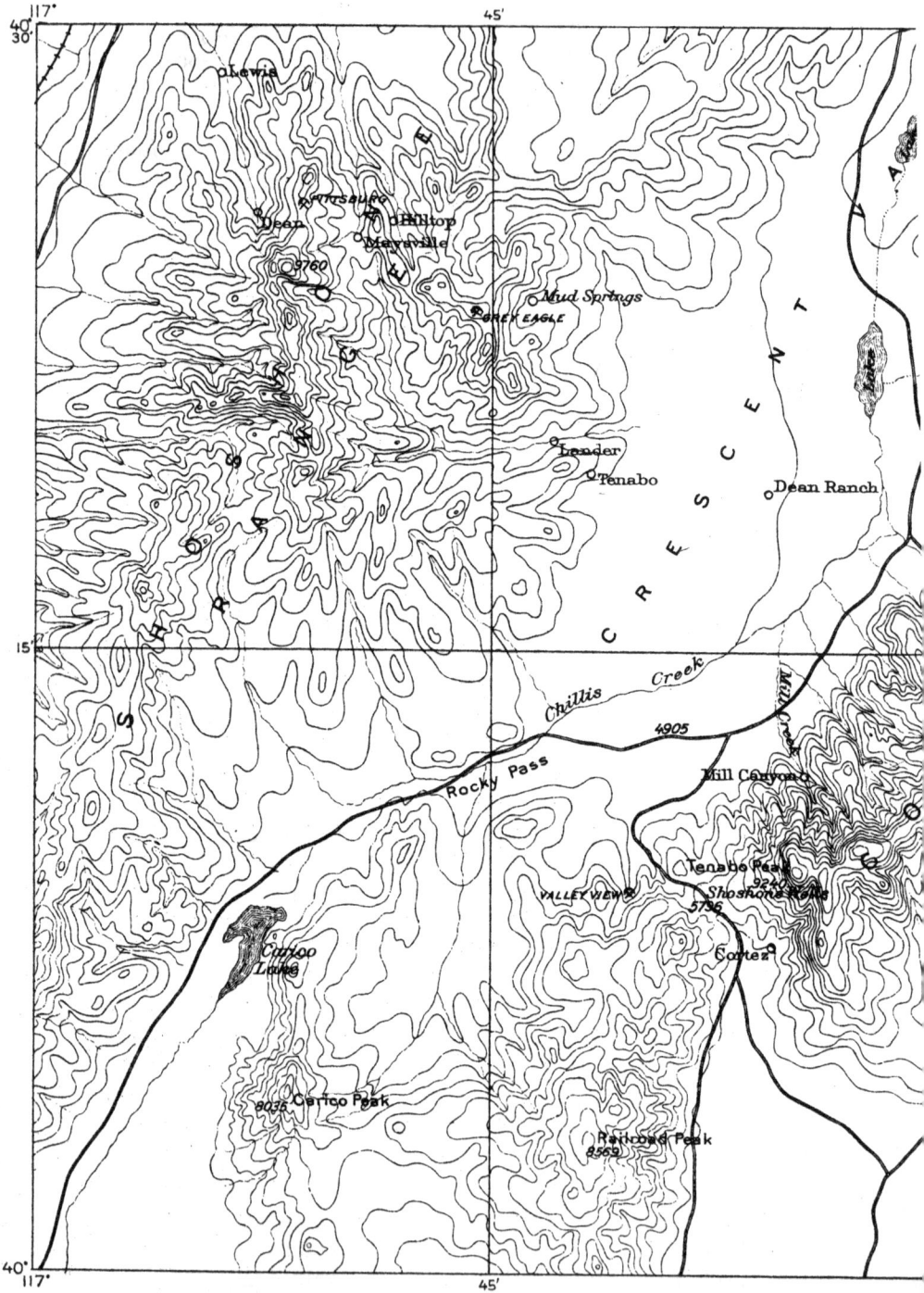

Topography from U.S. Geological
Exploration of the Fortieth Parallel
Clarence King, Geologist in charge

TOPOGRAPHIC MAP OF SOUTHERN Q

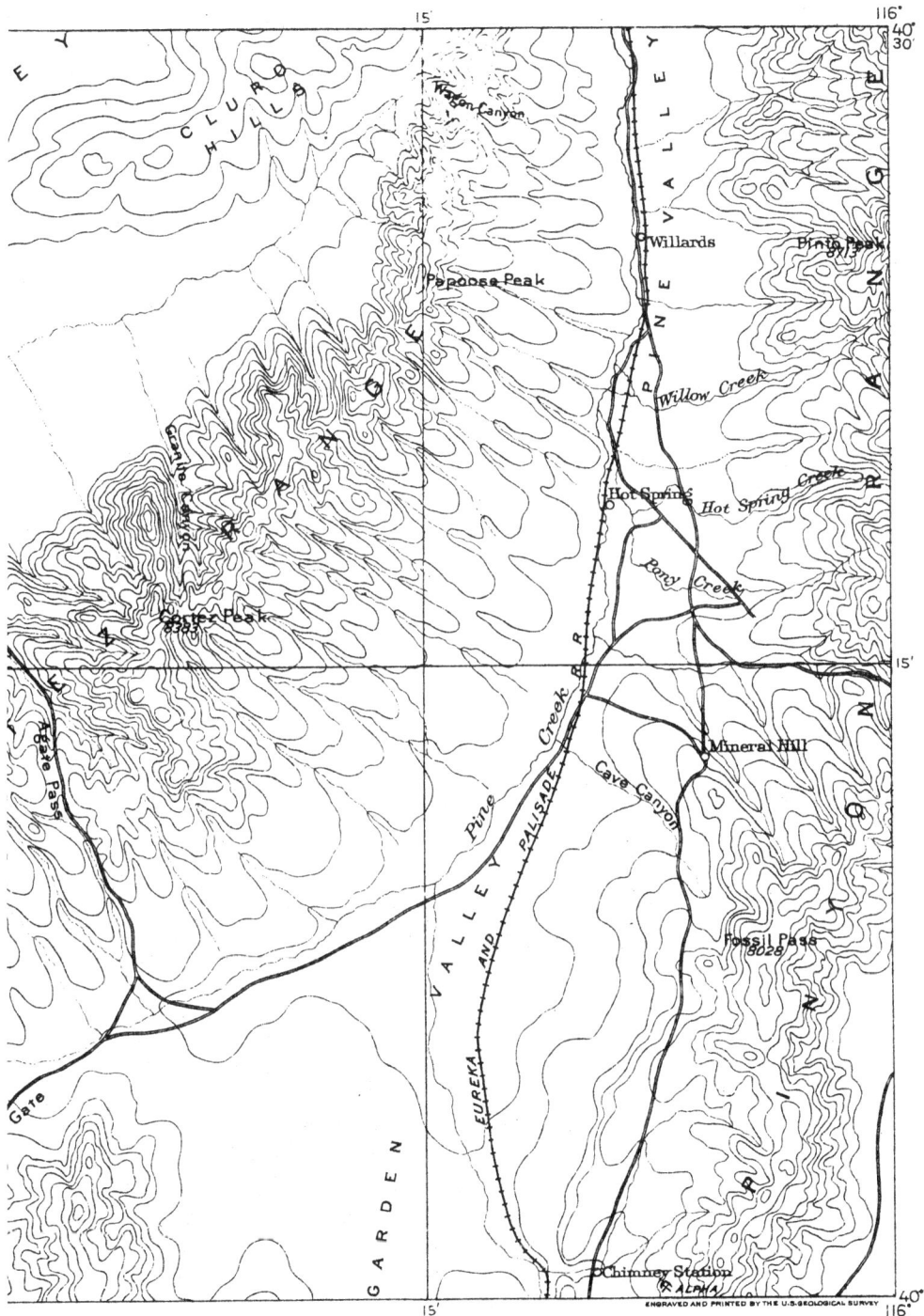

ER OF AREA SHOWN IN FIGURE 2

10 Miles

is shattered rhyolite, which is cut by many veinlets of iron-stained quartz that pans gold freely. Some of the placers were probably derived from these deposits.

A large acreage of ground west of Tuscarora has been located and sampled with drills. It is said that much of this ground will pay to work with dredges, and two companies are planning such operations. A large number of samples are reported to have given an average of about 14 cents per cubic yard.

FALCON MINE.

West of Tuscarora there is a great area of mountainous country, the higher peaks reaching elevations of 8,000 to 9,000 feet. The rocks are in the main Carboniferous quartzites capped by rhyolite and intruded by andesite and related rocks. A large mass of grano-diorite, probably older than the rhyolite and porphyry but intrusive in the sedimentary rocks, is exposed at the headwaters of Willow Creek and Rock Creek.

The Falcon mine, at the head of a small tributary of Rock Creek, is about 12 miles by wagon road west of Tuscarora. The mine was worked from 1879 to 1881 and the ore was hauled to Tuscarora. In 1884 a four-pan silver mill was built, but this was not operated and is now in ruins. The deposit is a fissure vein from 2 to 5 feet wide and is approximately vertical. Two deep shafts are sunk on it and shallow pits are dug at several places. The country rock is andesite, which near the vein is altered to a light-gray rock composed largely of white mica, but the fresh dark andesite is exposed at several places within 300 or 400 feet of the vein. The ore is highly siliceous and contains a small proportion of finely divided pyrite and other dark sulphides, which are banded with the quartz and show comb and rib-bon structure. The values are said to have been in ruby silver.

CORNUCOPIA.

The mines at Cornucopia, about 8 miles southwest of the stage sta-tion on Deep Creek, were operated actively in the seventies, when they produced, it is said, over a million dollars in silver. The ore was treated by pan amalgamation in a 20-stamp mill at Mill City, 2 miles below the town. The principal mines are the Leopard and the Panther, which were operated through shafts. The Leopard shaft is said to be 800 feet deep. When the camp was visited in 1908 all the deep workings were caved and only some shallow pits and sur-face stopes were accessible.

The country surrounding the Cornucopia district is a large area of low hills, which in the main are capped with rhyolite and obsidian. Under the microscope the denser rhyolite is seen to be composed of a glassy microlitic groundmass which contains phenocrysts of quartz,

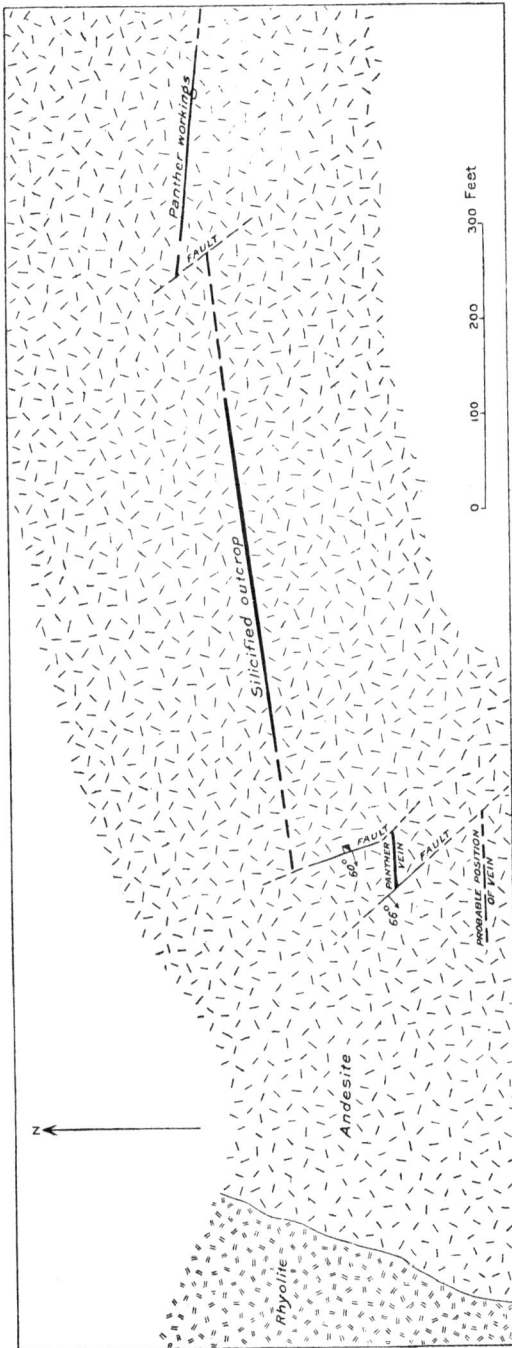

FIGURE 5.—Geologic sketch map showing Panther vein, Cornucopia district.

oligoclase, pyroxene, magnetite, and a little hornblende. It is more basic than most rhyolites of this area and approaches andesite in composition. Some phases of rhyolite are vesicular and some are pumiceous. The rhyolite is cut by intrusive andesite, to which the accessible ore deposits are restricted. An exposure of the andesite in a relatively fresh condition may be observed near the ruins of a stone house at Cornucopia. The rock is a dark porphyry, composed of a glassy groundmass which contains many phenocrysts of basic andesine and pyroxene, with large chloritic patches that seem to have resulted from the decomposition of hornblende. At the exposure named it may be noted grading into a highly decomposed sericitic phase of the andesite, which is white or stained with yellowish-brown iron oxide. Everywhere in the vicinity of the ore deposits the andesite is similarly decomposed. Masses of quartz porphyry occur in the area of the andesite and are probably intruded into it. The quartz porphyry is light colored and is composed of a groundmass, presumably microcrystalline, which contains white mica, feldspars, and small phenocrysts of resorbed quartz. The rock outcrops in the tunnel of the Leopard mine, but none of the deposits, so far as known, is inclosed in it.

The ore deposits are sheeted zones in decomposed andesite. The ore is white quartz, which carries a very small proportion of dark sulphides, forming narrow ribbons in the quartz. Pyrite, argentite, and gray copper are present, and ruby silver is said to have been an important ore mineral. On the surface the ore minerals are mainly horn silver and a yellow mineral which is probably pyromorphite. The proportion of the sulphides present is very small, but they must have been rich, for the ore is said to have carried 400 ounces of silver to the ton for mill runs. In some of the ore the minerals are arranged symmetrically with respect to the walls, the quartz crystals pointing to the center of a druse, showing that the ore was deposited in open spaces. The country rock along the veins is, however, silicified and otherwise altered by the vein-forming solutions, and at some places carried workable values.

The Panther vein in the principal workings southwest of the silicified outcrop shown in figure 5 strikes S. 78° W. and dips 83° N. to 90°. Here underhand stoping has been carried down for a distance of 60 feet along the strike. The country rock is altered andesite, a soft kaolinized mass cut through by veinlets of white quartz. At the west end of the stope a smooth slickensided fault strikes northwestward, cutting off the vein. The surface of the fault shows striæ inclined northwestward to a line along the direction of steepest dip and making an angle of 15° with it. The country to the north has been prospected for the vein, but it has not been discovered in that direction on the southwest side of the fault. If the fault is normal, the vein to the west of it should be found south of the present workings, where possibly it is represented by some poorly defined masses of quartz which outcrop at that place.

A few rods northeast of the surface stope is a prominent ledge which strikes a few degrees north of east. This ledge, which is several hundred feet long, is the altered andesite somewhat silicified, fractured, and seamed with quartz veinlets. To the east of the ledge a vein in altered andesite has been developed for a few feet and some ore has been stoped. This vein is probably the faulted continuation of the silicified outcrop of quartz which lies between the two groups of workings.

GOOD HOPE DISTRICT.

In the Good Hope district, which is about 12 miles southwest of Cornucopia, mining was carried on in the early eighties, when the camp is said to have produced over $100,000 in silver. The principal mines are the Buckeye and Ohio, the Snyder, and the Page & Kelley. All these mines were inaccessible in 1908 except the Buckeye and Ohio, which was under water above the adit level. The deposits are in the main sheeted zones in rhyolite flow breccia, but altered andesite is exposed at several places and it is probably the country rock for some of the ore deposits. The leaching of the country rock is extensive near some of the deposits, but is not so general as at Cornucopia. Pyritization is, however, more pronounced in the wall rock and the ore at Good Hope contains a much greater amount of the sulphides. The Buckeye and Ohio mine is on Fourmile Creek near its junction with Atlantic Cable Gulch. The mine was operated from 1882 to 1884, and the ore was run through a 5-stamp mill 1½ miles below the mine. This mill was equipped with roaster, pans, and settlers and employed the Reese River process. A small concentrator was built in 1903, but the treatment employed was presumably unsuccessful, as only a small amount of ore was put through it.

On the surface above the adit level obsidian and other varieties of glassy lavas outcrop at several places. In the mine the rhyolite is a light-colored flow breccia, which is locally altered to a white claylike mass. Along a zone of movement it has been converted' to gouge.

The lode is composed of veinlets of quartz and sulphides, which include masses of the country rock highly altered and partly silicified. A tunnel driven on the vein for 300 feet southwestward gives a depth below the surface of about 65 feet. At some places stopes are carried to the grass roots and several winzes are sunk upon the vein. The ore is composed of quartz, pyrite, arsenopyrite, freibergite, stibnite, and dark ruby silver.

117°
41°
45'

Warm Spring
6859

ROCK VALLEY TRAIL

ROCK CREEK
SQUAW CREEK
Rock Creek VALLEY

S H O S H O N E

M E S A

R Y

Rock Creek

M E A D

45'

Stony Point
6850

Humboldt River

Shoshon
4665

Argenta
4575

Battle Mountain

NEVADA CENTRAL R.R.

S H O S H O N E

R A N G E

Beowaw
4717

WHIRLWIND VALLEY

40'
30'
117°
45'

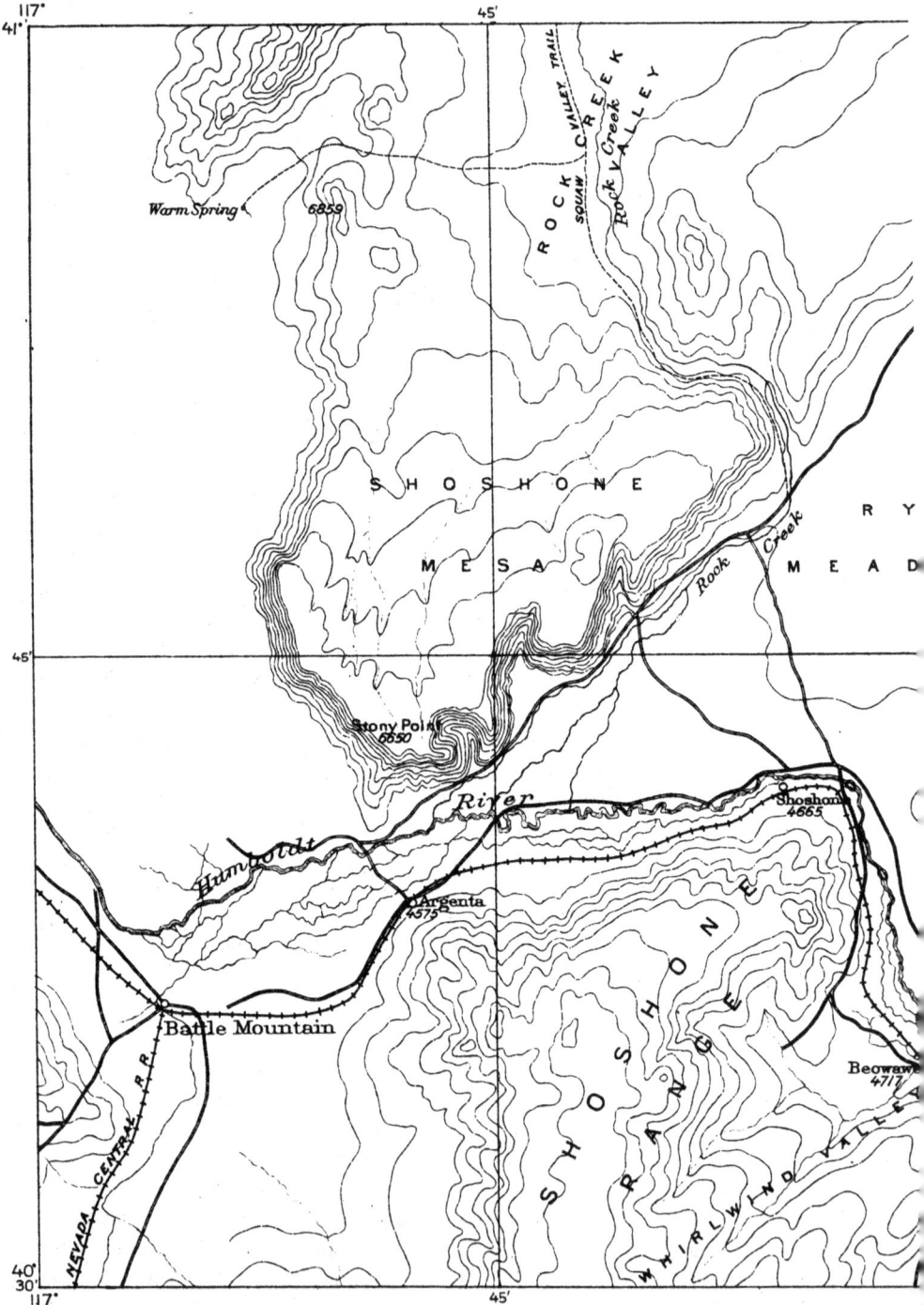

Topography from U.S. Geological
Exploration of the Fortieth Parallel
Clarence King, Geologist in charge

TOPOGRAPHIC MAP OF SOUTH-CENTRA

10   5

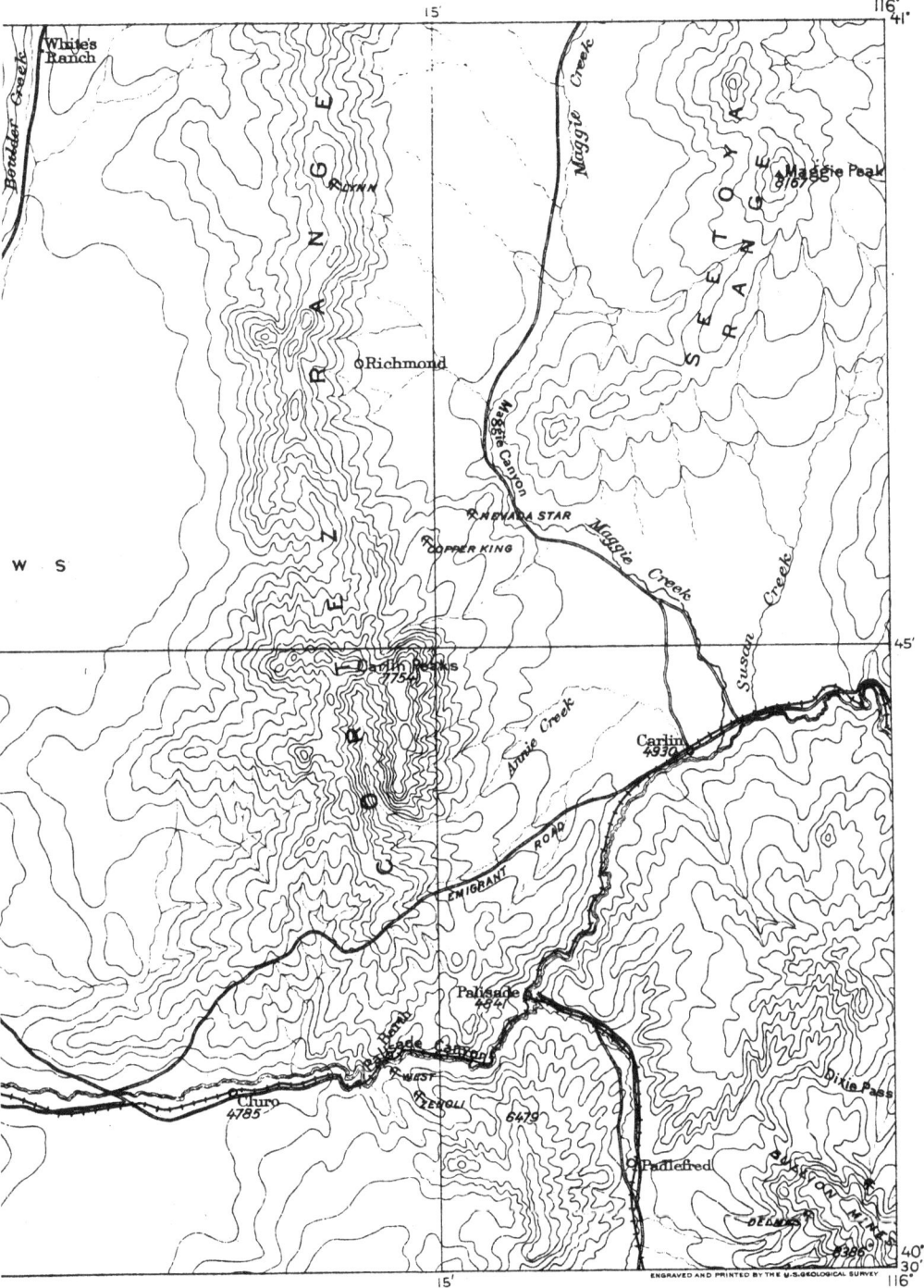

UARTER OF AREA SHOWN IN FIGURE 2

10 Miles

The vein, as shown in figure 6, is followed by the adit for a distance of 300 feet. At the breast, where it dips steeply westward, it is cut off by a fault which strikes 37° E. This fault follows a vein also and carries much crushed ore and some banded ribbons of unbroken ore. A stope from this vein is carried to the surface. Near the junction of the two veins a second fault, which deviates from the principal fault about 5°, cuts off the west vein. The ore therefore forms a letter V, which points to the south. If the faults are normal the continuation of the principal vein will be found south of the present workings, on the west side of the fault.

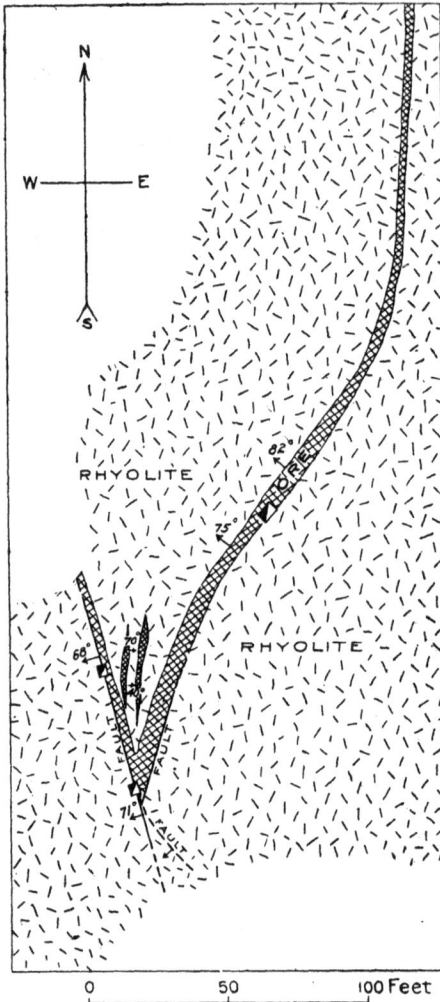

FIGURE 6.—Plan of adit level of Buckeye and Ohio mine, Good Hope district.

Near the head of Atlantic Cable Gulch, about 1 mile above the Buckeye and Ohio mine, several prospects show brecciated rhyolite silicified and cemented by quartz and dark sulphides. One of the lodes outcropping boldly on either side of the gulch strikes N. 10° E. and dips 35° W., and its outcrop forms a V pointing up the gulch. In places it carries a considerable quantity of dark pyritic ore which has been crushed and recemented by white and barren-looking quartz. At one place an abandoned incline is driven on the lode to a depth of about 30 feet, exposing a considerable mass of quartz and sulphides.

BURNER HILLS.

GENERAL FEATURES.

The Burner Hills, which are some 10 miles west of Good Hope, rise about 800 feet above the broad undulating plain of the Owyhee Desert, which lies to the east and north. From Good Hope to the

Burner Hills this plain is covered by beds of rhyolite and rhyolite pumice. Nearly everywhere these beds are flat or dip at low angles in various directions, but as they approach the hills the marginal fringe of rhyolitic pumice becomes steeply upturned and dips away on the east side at angles up to 35°. Still higher on the hills siliceous shales with beds of intercalated limestone outcrop at many places. These beds are highly tilted and show a considerable variety of attitudes, but the prevailing dip appears to be away from the central axis of the hills, the summit of which is a fresh andesite showing massive columnar jointing. The andesite intrudes the sedimentary rocks and probably rhyolites also, although the contact at this place was not seen. A number of claims are located on these hills, but the Mint mine is the only one on which any considerable amount of work has been done. This mine was operated in the early eighties and shipped about $30,000 worth of lead-silver ore to smelters. Active operations were suspended in 1893, and since that time but little work has been done.

MINT MINE.

At the Mint mine a tunnel is driven southwestward for 175 feet to the lode, which it follows for 300 feet. The lode strikes S. 25° W. and is approximately vertical. Here and there stopes have been carried upward and a winze is driven on the lode below the adit level. The ore consists of galena, sphalerite, pyrite, arsenopyrite, and chalcopyrite, in a gangue of quartz and calcite. Near the surface lead carbonate and iron oxide are present. The sulphides and quartz occur as ribbons parallel to the walls or as masses impregnating the andesite, which is somewhat altered by the vein-forming solutions. The high-grade ore is said to be irregular and bunchy in the vein, but a zone up to 4 feet wide is regarded by the owners as available for concentrating, as a considerable proportion of the silver values is in galena.

South of the Mint mine are several small veins of iron-stained siliceous silver ore. Some of these cut across the sedimentary rocks; others occur as stringers parallel to the bedding.

CENTENNIAL RANGE.

GENERAL FEATURES.

The Centennial Range, which lies between Deep Creek on the south and East Fork of the Owyhee on the northeast, is about 20 miles long and from 5 to 10 miles wide, its principal axis trending northeastward from Lime Mountain to Montana City. The higher summits rise from 3,000 to 4,000 feet above the Chellis Valley, which topographically is a part of the Owyhee Desert, a great rolling plain of sagebrush that extends northwestward far into southern Idaho.

The range is separated from the Jack Creek Mountains, to the southeast, by Bull Run Basin, which is drained by Bull Run Creek. This stream flows westward through a deep V-shaped canyon that separates the central mountain mass of the Centennial Range from a narrow, lofty ridge which extends southward toward Deep Creek. The highest part of the range is a compact group of mountains that lie between Bull Run Creek and Blue Jacket Canyon and cluster about Porter Peak, the loftiest summit. North of Blue Jacket Canyon the hilly country extends to Mountain City and beyond that northward into Idaho. The topographic expression of the range is due to faulting modified by erosion and to a trivial extent by glaciation.

The rocks are mainly Paleozoic quartzites, limestone, and shales. On the slopes of Porter Peak the prevailing dip is northward, almost at right angles to the principal axis of the mountains. The great faults on which the chief structural features depend have so broken the geologic column that in the absence of fossils it was not possible to determine the age of the rocks satisfactorily in the limited time devoted to the work, but from the lithologic descriptions published by the King Survey for the country to the south it seems highly probable that the Carboniferous formations have the widest distribution. The great quartzite beds which form the southern portion of Porter Peak and which include the ore deposits at Edgemont and at Bull Run are regarded as Carboniferous. This is a medium-grained quartzite, which through great thicknesses shows comparatively slight variations. At the Bull Run mine there are some thin-bedded siliceous, shaly layers, and on the north slope of the hill south of Edgemont some fine conglomerates were noted, but the great mass of the formation is a dull gray or pink quartzite, massive, thick bedded, and strongly jointed, at many places showing too little evidence of stratification to define its attitude. On the ridge north of the stream which flows westward from Porter Peak through Edgemont the quartzite is overlain by a great series of limestones. This series is several thousand feet thick and on the north slope of Porter Peak grades into black shales which still farther north are overlain by a thick series of limestones and shales. The great expanse of hilly country between White Rock and Mountain City was not traversed in this reconnaissance, but at a distance it appears to be composed in the main of sedimentary rocks capped with rhyolite.

At a number of places, including Blue Jacket Canyon on the southeast side of the range and the ridge south of White Rock Canyon on the northwest side, the limestones are intruded by medium-grained granitic rocks. None of the outcrops of these intrusive masses are very extensive, but at the north end of the range they are numerous, and possibly they are connected in depth. A characteristic specimen from one of the larger masses on the ridge south of White Rock Canyon is a granodiorite composed of oligoclase, quartz, orthoclase, biotite, and hornblende. No garnetization of limestone was noted near these intrusions. At Mountain City the sedimentary rocks are intruded by a granodiorite, which is more coarsely crystalline and carries more quartz and orthoclase than the average granodiorite. North of Aura and also on California Hill, 2 miles above Mountain City, garnet, epidote, tremolite, and actinolite have formed extensively near the contact of granodiorite with limestone.

Here and there in the Bull Run Basin, protruding through the cover of granite, are outcrops of rhyolite and basalt, and in the country to the east, extending to a great but unknown distance, are thick beds of rhyolite with a subordinate amount of rhyolite flow breccia. Very extensive beds of rhyolite occur also in the lower country to the west and south of the Centennial Range. This rhyolite is much younger than the sedimentary rocks composing the main mass of the Centennial Range, and if it were in its original position it would now be above the sedimentary rocks instead of forming the floors of the lowest depressions, such as Bull Run Basin. The present structure was brought about by faulting and tilting. North of Bull Run Creek, on the south slope of the high mountains which form the central portion of the range and near the trail from Aura to the Bull Run mine, there are bluffs of volcanic agglomerates and tuffs composed of rhyolite fragments with a large proportion of diorite porphyry. At the base of the exposed portions of the beds there are some layers of shaly coal. The agglomerate beds are of Tertiary age, and as they dip toward the Paleozoic rocks which form the central and most elevated portion of the range, there must be a fault of several thousand feet between the two systems of rocks.

ORE DEPOSITS.

In 1869 a party composed of Cope, Dixon, and others, going from Silver City, Idaho, to White Pine, Nev., made discoveries and located claims near Columbia and in Blue Jacket Canyon near by. These discoveries attracted considerable attention to the district, and in the early seventies silver mills were built in Blue Jacket Canyon and at Mountain City. In 1875 Edward Stokes built a mill at Columbia to treat the ore from the Revenue, California, and Infidel mines. All these mills employed the Washoe or the Reese River process, and the conditions for operation were most trying, as it was necessary to haul the bulky supplies required for silver milling 75 miles or more by wagon road. Although a considerable tonnage of chloride ore, taken from the upper parts of the deposits, was put through these mills, it is a question whether much profit was made from the operations. Owing to the insuperable difficulties in

117°
45'
41° 28
41° 15'
41° 117°
45'

Rose Mtn
7949

Midas

BLUFFS

Y H E E

SQUAW VALLEY

ROAD

Sunset Gap

Lake
5606

SQUAW VALLEY TRAIL

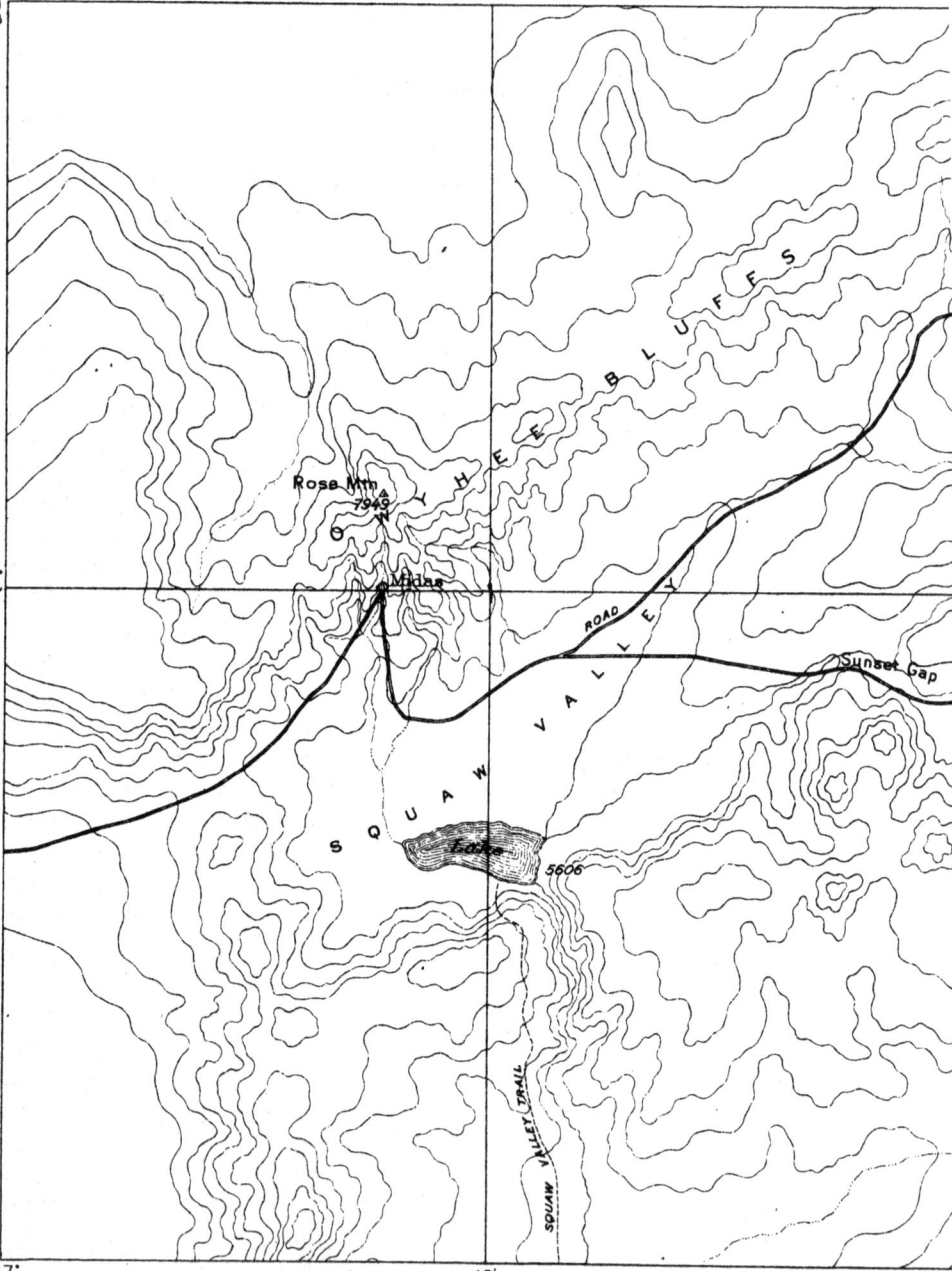

Topography from U.S. Geological
Exploration of the Fortieth Parallel
Clarence King, Geologist in charge

TOPOGRAPHIC MAP OF NORTH-CENTRAL

10   5

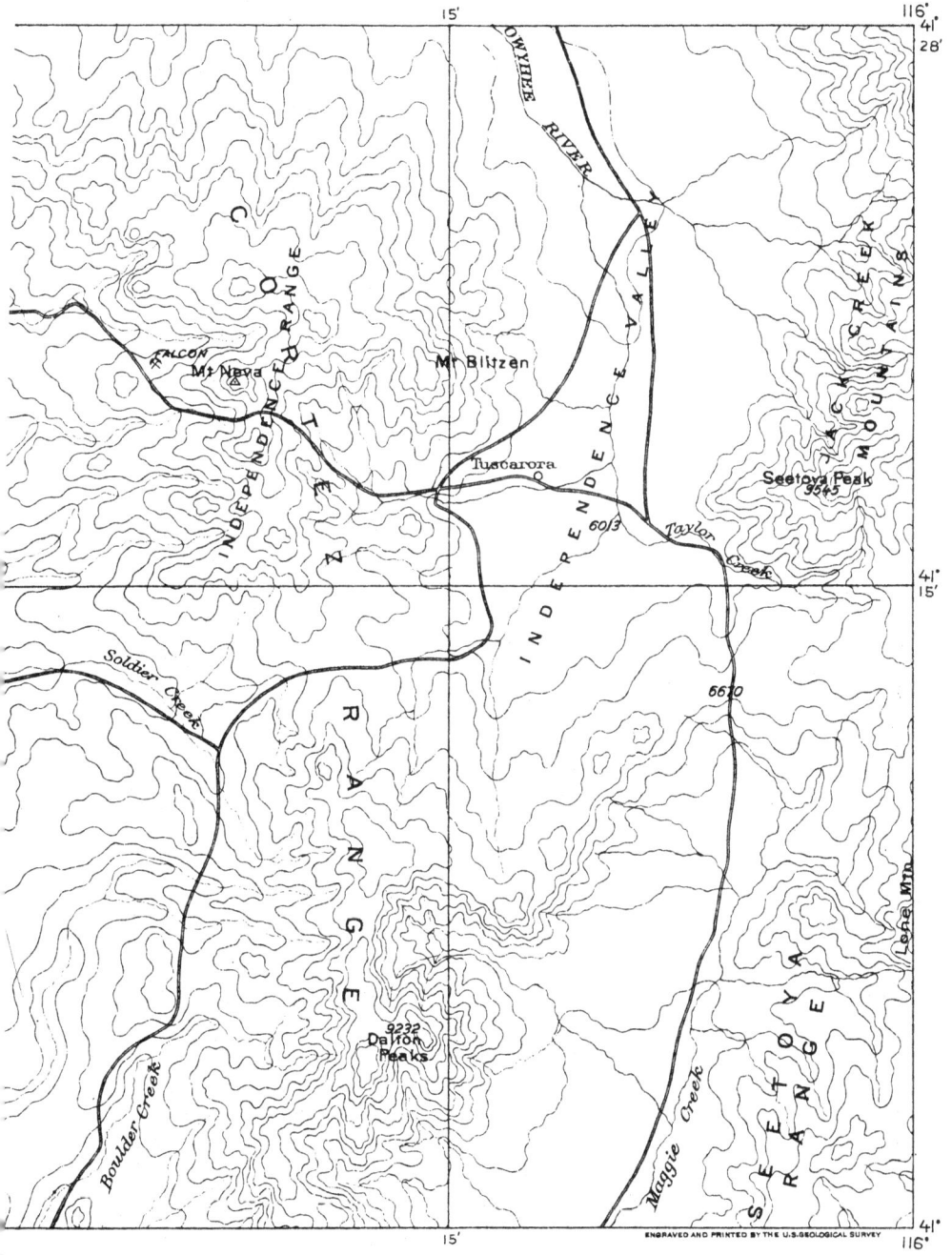

UARTER OF AREA SHOWN IN FIGURE 2

10 Miles

the way of getting cheap transportation, together with the faulted condition of the lodes, the operators lost heart and turned their attention to other fields. In the nineties the mining industry was revived by the discoveries of gold ores at Edgemont, and in 1906 the country again felt the stimulus of enthusiasm for prospecting which swept over Nevada from the southwest. In that year several gold veins were discovered near Aura and elaborate preparations were made to reopen some of the silver mines at Mountain City, but these operations were stopped in 1907 when it became difficult to obtain money for prospecting or development work.

The ore bodies are fissure veins which cut across the bedding of sedimentary rocks, bedding-plane deposits which follow the stratification, and fissure veins in granite. The sulphide ores fall into two general classes—(a) gold deposits of highly siliceous ore carrying a small percentage of pyrite and galena and (b) silver deposits carrying these minerals in greater abundance, together with a small proportion of arsenic and antimony minerals. At Edgemont and Aura the deposits are in the sedimentary rocks, but at Mountain City they are mainly in the granodiorite. They were formed before the faulting took place. In every mine where any considerable amount of development work has been done faults have been encountered. These are nearly everywhere of the normal type, which implies a downward movement of the hanging wall.

The rhyolites that flank the mountains are probably younger than the deposits which have been developed in the Centennial Range. At Gold Circle, Lynn, and elsewhere this rhyolite carries gold deposits that are of a different type from those of this range, and there is no reason why it should be avoided by the prospector, especially where it is intruded by dark rocks (andesite) and in areas where it is leached white by hot waters.

LIME MOUNTAIN.

Lime Mountain forms the southern extremity of an elevated ridge about 6 miles long, which lies between Bull Run Basin and Chellis Valley. This ridge, which may be regarded as a southward extension of the Centennial Range, is separated from it by the very steep canyon occupied by Bull Run Creek. The rocks at Lime Mountain are in the main dark-gray limestones, which at the summit of the mountain dip from 20° to 40° W. The limestone is cut by dikes and other intrusions of igneous rocks, which include quartz porphyry, andesite, and diabase. Locally the limestone is metamorphosed to a coarse-grained marble, but the metamorphic action is not intense and very little garnet or hornblende has been developed. The principal workings are at the Eldorado mine, where a 200-foot tunnel driven westward into the mountain is connected with a shaft 115 feet deep. At their intersection and upward to the surface much of the workings is in ore that is said to average several per cent of copper, with values in gold and silver. The ore consists of pyrite, chalcopyrite, and bornite, which are intergrown with white and black mica, calcite, and quartz. The deposit, which is probably of contact metamorphic origin, has been fissured since deposition, and there has been some secondary chalcocite enrichment of the copper-iron sulphides. The rock that caused the metamorphism could not be determined. About 1,000 feet northwest of the mine is a considerable area of quartz porphyry, and a diabase dike, outcropping on the crest of the hill, 500 feet above the mine, strikes toward it. On the crest of the ridge, at places which are much nearer to those rocks than the Eldorado, the limestone is not greatly metamorphosed, except in some localities where it is a very coarse marble.

COLUMBIA AND AURA.

GENERAL STATEMENT.

Columbia, which is situated at the north end of Bull Run Basin, was in the seventies the headquarters for prospecting and mining in the central part of the Centennial Mountains. In the boom times of 1906 Aura was founded a mile or two below Columbia, and it is now the post-office and supply point for the east side of the range.

Several mines are located near Columbia, but all of them except the Big Four have long been idle. This mine, which was discovered in recent years, made several shipments of ore in 1908 and is commonly regarded as a property of some promise. The silver mines which were abandoned years ago were relocated recently, but little work has been done except enough to hold the ground.

In the canyon of Blue Jacket Creek, which joins Columbia Creek in the north end of Bull Run Basin, there are several mines and prospects, but none of them have been extensively exploited except the Blue Jacket, which was worked in the seventies. Two 5-stamp amalgamation mills were built in Blue Jacket Canyon to work the gold ores, utilizing the water power of Blue Jacket Creek. Both were idle when the camp was visited in 1908. Some years ago a company was formed to mine the gravels of Bull Run Basin, which are said to carry gold, and considerable money was spent in ditches, flumes, and pipe. The effort was presumably not successful, and the project was abandoned after a few hundred yards of material had been put through the sluice boxes.

MINE DESCRIPTIONS.

Infidel mine.—The Infidel mine, which is located on the hill above Columbia, was worked from three tunnels driven northward into the hill. The mine has been idle for many years and the middle tunnel only was accessible in August, 1908. The rocks near by are mainly limestones and shales, but between the Infidel and Big Four mines is a body of granodiorite of considerable size. The lode strikes 30° W. and lies approximately parallel to the bedding of the country rock, a dark shaly limestone. It is a sheeted zone composed of three or four veinlets of white quartz closely spaced and nearly parallel. The ore is white quartz, containing a very small proportion of galena, pyrite, and other dark sulphides, said to carry high values in silver.

Big Four mine.—The Big Four mine is on the east bank of Columbia Creek, about one-fourth of a mile above Columbia. The deposit is a flat vein which follows the bedding of a shaly limestone dipping from 10° to 15° S. Three tunnels, each about 100 feet long, are driven on the vein, exposing an ore body that has a maximum width of about 5 feet. The lode is a sheeted zone in the shaly limestone, and the ore is composed of quartz, calcite, pyrite, galena, zinc blende, and chalcopyrite, with iron oxide and some green copper carbonate. It is said to carry high values in both silver and gold, the cobbed ore

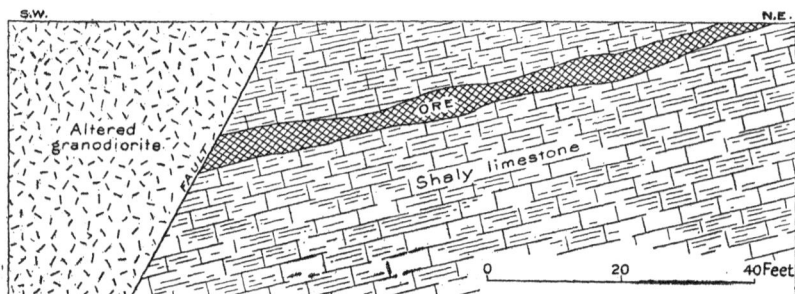

FIGURE 7.—Cross section of Big Four lode, Columbia.

being rich enough to pay for shipment. Near the lower tunnel the vein and the inclosing limestone are in faulted contact with granodiorite, as shown by figure 7.

Columbia Queen mine.—The Columbia Queen mine, formerly the Bonanza, is on the west side of Columbia Creek, about 300 yards west of the Big Four mine. Like the Infidel and Revenue, it was the property of the Stokes Company, which operated it in 1875. The deposit, which resembles that of the Infidel, is a sheeted zone in limestone. The lode is 3 feet wide and carries galena, zinc blende, pyrite, and gray copper, with high values in silver and some gold. As shown by figure 8, the deposit dips toward the south and is displaced by three northward-striking faults, each of which offsets the lode toward the north.

Blue Jacket mine.—The Blue Jacket mine is located at the head of Blue Jacket Creek near the divide between this stream and Silver Creek. The deposit was worked through two tunnels and a deep shaft, all of which were inaccessible when the mine was visited. The country rock is limestone and, to judge from the dump, the ore is similar to other silver deposits near by. It is composed of white quartz carrying a small percentage of galena, pyrite, and zinc blende, with copper carbonate, iron oxide, and other minerals. The ore was carried over a wire tramway to a 20-stamp pan-amalgamation mill, the ruins of which may be seen below the mine at a bend of Blue Jacket Canyon.

Jack Pot mine.—The Jack Pot mine is on the south side of Blue Jacket Canyon, about 300 yards below the lower tunnel of the Blue Jacket mine. It was discovered in 1906 by Blewett Brothers, who have done about 800 feet of development work. The country rock is

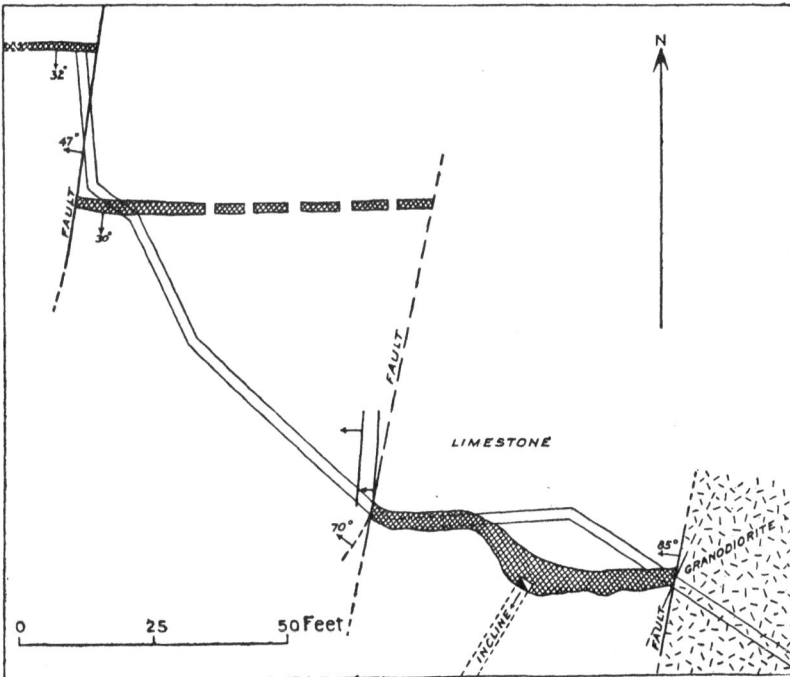

FIGURE 8.—Plan of tunnel level, Columbia Queen mine, Columbia.

limestone, which, some 200 yards south of the mine, is intruded by a broad dike of granodiorite. The lode, which is developed by four tunnels driven one above another, strikes northwestward and dips 35° SW. and carries shoots of ore which have a maximum width of about 5 feet. The ore is in the main decomposed, silicified limestone, stained with iron oxide. Galena, pyrite, chalcopyrite, and zinc blende are found in the lower workings. About 60 tons of the ore has been put through the Walker mill, but the saving by amalgamation was not satisfactory. A considerable tonnage on the dump and in the mine is said to be rich enough in gold to yield a profit on cyaniding.

IDAHO
NEVADA

42° 117°

116°

N

Little Owyhee R.

Owyhee

COUNTY
COUNTY

HUMBOLDT
ELKO

O W Y H E E   D E S E R T

Milligan   Creek

Winters

Creek

BURNER MINE

41°
30°

117°

116° 3

TOPOGRAPHIC MAP OF NORTHERN C

10

5

TER OF AREA SHOWN IN FIGURE 2

10 Miles

California mine.—The California mine is located on the east slope of Porter Peak, near the head of a small tributary to Blue Jacket Creek. The lode is a siliceous replacement vein, which cuts across the bedding planes of limestone and dips southwestward at a high angle. The ore consists of white quartz, galena, and iron oxides, with some green copper carbonate. Where exposed in the principal working pit the lode is 4 or 5 feet wide.

Tiger lode.—The Tiger lode is on the north slope of Porter Peak, at the head of Silver Gulch, which drains westward from the central axis of the range. The lode dips 60° W. and is opened along the surface by five short tunnels driven one above another. Some ore from this mine is said to have been packed by mules to Mountain City in the seventies. The lode is a 3-foot vein of ore, very high in silica, and contains galena, pyrite, and a small amount of zinc blende. The surface ore carries copper carbonate, pyromorphite, and iron oxides and is said to be rich in silver. North of the Tiger lode, at the head of Silver Creek, there are a number of abandoned· shafts that were driven on lodes of siliceous ores in limestone. Northwest of these shafts, on the ridge between White Rock and Silver Creek, dark limestones dip from 45° N. to 90°. Three or four silver veins striking northeastward cut across the limestone, and on the crest of the ridge these veins outcrop boldly at several places. With better transportation facilities these deposits should be regarded as worthy of more careful prospecting.

Humboldt mine.—The Humboldt mine is at the head of Polaris Gulch, about half a mile southeast of the Blue Jacket mine. The country rock is contorted shaly limestone, and the vein, which is 1½ to 3 feet wide, is composed of white quartz, galena, pyrite, chalcopyrite, and gray copper. An incline is driven southeastward at a low dip, but when the mine was visited most of it was under water. About 100 yards east of the Humboldt shaft a surface pit in granodiorite shows a strongly sheeted and leached zone, which is said to carry up to $3 a ton in gold.

Polaris mine.—The Polaris mine is in Polaris Gulch, about one-fourth mile southeast of the Humboldt. An incline is sunk about 200 feet deep and from it two levels are turned. A fissure dipping 32° SE. cuts across the bedding of the limestone. Here and there along the fissure are masses of quartz and silver-bearing sulphides. This mine supplied some ore to the Columbia mill in the seventies.

Aura King mine.—The Aura King mine is in Blue Jacket Gulch, between the Walker and the Aura King mills. The vein, which is in limestone, dips 30° SW. and has been followed on the strike for 200 feet. It is from 3 inches to 1 foot wide and is said to carry high values in gold.

EDGEMONT.

Lucky Girl group.—Edgemont is an active little mining camp located on a branch of White Rock Creek, on the west slope of the Centennial Range. It is supported in the main by the mining and milling operations of the Montana Gold Mining Company, owning the Lucky Girl group of claims, which includes all of the deposits which have been extensively developed on the west side of the range. These claims were purchased in 1898 by Alex. Burrell, and a 20-stamp mill was built in 1902 and has been in continuous successful operation ever since, treating about 60 tons of ore a day. Electric power is transmitted from a plant installed on a tributary of White Rock Creek, 4 miles north of Edgemont. The mines of this company comprise about 5 miles of underground workings and extensive ore bodies have been developed. The deposits include the Lucky Boy, Lucky Girl, and Big Bob veins. The Lucky Boy vein is the most extensively developed and has supplied most of the ore to the mill. This group of mines and the Bull Run mine on the south slope of the mountain have yielded since they were opened about $1,000,000, chiefly gold.

As now arranged, the ore is dropped to the fifth-level adit and is drawn by mules to the portal, where it is fed automatically to a bucket tramway, 3,600 feet long, which carries it to the mill 800 feet below. At the mill the ore falls on grizzlies and the large rock goes to a 9 by 11 Gates crusher. There are four batteries of five stamps each. The fine rock goes to the outside stamps, which weigh 750 pounds, and the coarse rock to the inside stamps, which weigh 1,000 pounds. The stamps drop 8 inches eighty-five times a minute. About 50 per cent of the values are caught on amalgamation plates, from which the pulp passes to three Wilfley tables and one Pinder concentrator, where the galena and pyrite are removed. To utilize a water supply from a gulch north of the mill, the tailings are passed through a flume to a dam about 2,000 feet below. Here the slimes are drained from the sands and the sands are loaded into nine 50-ton cyanide tanks. The slimes are collected at a second dam below and are loaded dry into a mixer, where they are broken and mixed with water to the desired consistency. From this they pass to the agitators and thence to decantation tanks. The total extraction is from 90 to 95 per cent, the ore carrying from $5 to $10 in gold and 1 ounce of silver to the ton. The ore yields about 0.6 per cent in concentrates that carry about $115 in gold, silver, and lead.

The deposits are fissure veins in siliceous sedimentary rocks which are steeply tilted, folded, and faulted. The country rock is a brown or gray quartzite of rather uniform grain, with here and there thin beds of siliceous shale, which at some places is dark gray or nearly black.   Near the veins a little sericite or white mica has been developed in the quartzite, together with a small amount of pyrite, but, compared with deposits in the porphyries, the wall rock is but slightly changed in composition.   Between the shaft of the Lucky Girl and the portal of the low-level tunnel, conglomerate beds are interstratified with the quartzite.   These beds are only a few feet thick and consist in the main of well-rounded pebbles of uniform size, about one-half inch in diameter.   Most of them are white quartz, with here and there a pebble of red jasper.   In the Lucky Boy mine small folds may be observed at several places on the surface and underground, and the difficulties of interpreting the structure are increased also by pronounced sheeting across the bedding in the same general direction as the lodes and by jointing in other directions.   The prevailing dip of the sedimentary rock is 30° or more northwestward, but at some places the dip is about 40° NE.   Some of the quartz was deposited before the deformation of the rocks was completed, as is shown by the

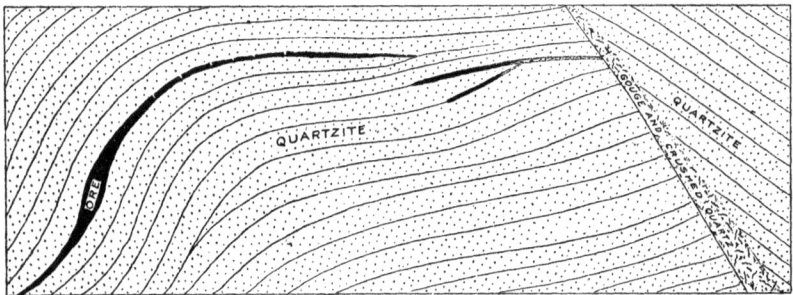

FIGURE 9.—Sketch of wall on level 8, 1,200 feet from portal, Lucky Boy mine, Edgemont.

fractured ribbons of quartz indicated in figure 9.   The principal deposits, however, are little affected by folding and their deposition must have followed the greater deforming movements.

The veins, which at most places cut across the bedding of the quartzite, outcrop plainly on the surface, where they are composed of white quartz slightly stained with iron oxide.   The three lodes that are most developed strike northeastward and dip from 15° to 45° SE. The ore is highly siliceous and carries more than 90 per cent of quartz. The sulphides are pyrite, galena, and arsenopyrite.   The gold is associated mainly with the sulphides or with their oxidation products. The oxidized and sulphide ores carry approximately the same values, from $5 to $10 a ton, and there is no evidence of secondary enrichment other than that due to the removal of soluble constituents; this, owing to the comparatively firm condition of the ore, is small.   Copper carbonates are rare or altogether lacking, and no zinc minerals were noted. Considerable fissuring with slickensiding must have taken place before deposition, as is indicated by the section shown in figure 10, where slickensided fissures stop at the vein, which they do not disturb or cross. Movement subsequent to deposition has produced much gouge along the walls, but this is barren except where it contains crushed quartz. The sharp contact of country rock and vein quartz

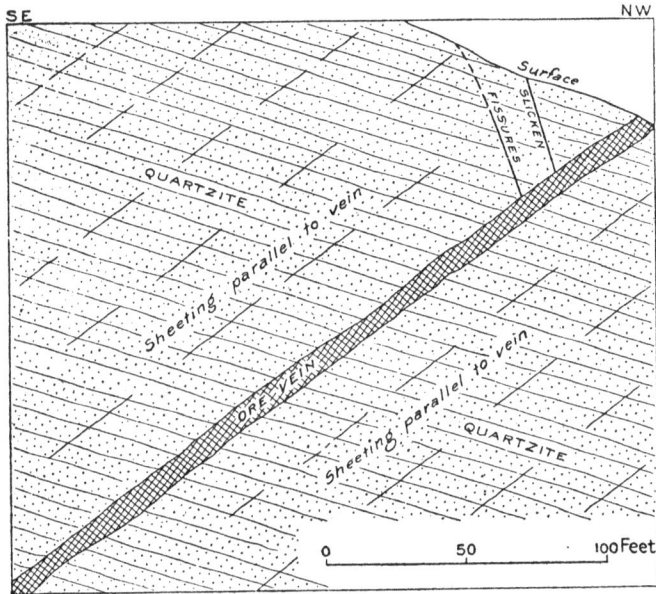

FIGURE 10.—Section S. 40° E. through Lucky Girl vein, Edgemont.

and the dependence of vein fillings on previous openings are illustrated by figure 11.

The Lucky Boy vein has been developed along the strike for a distance of about 3,200 feet and through a vertical depth of 400 feet. On the fifth level a fault which strikes approximately with the vein

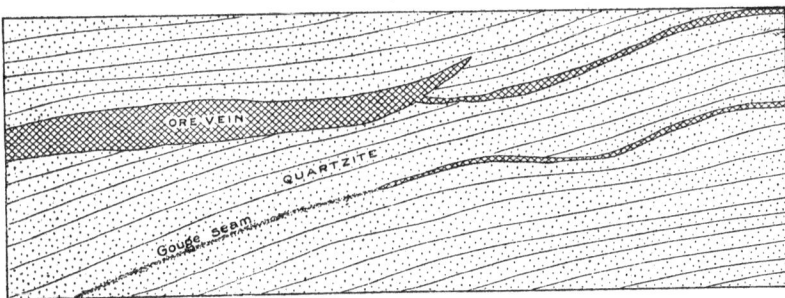

FIGURE 11.—Sketch of the northwest wall on level 8, Lucky Boy mine, Edgemont.

and dips in the opposite direction, from 40° to 80° NW., may be located for a distance of 2,000 feet, being exposed at five places in various drifts and raises. The throw of the fault is about 30 feet, measured vertically, as shown in the section given in figure 12. It

results in a duplication of the vein on level 5. At some places the
wide zone of faulting carries much gouge, and it is difficult to locate
the main fault plane precisely or to determine the direction of dip, but
in the main the slicken planes dip to the depressed side, indicating
that the fault is normal. In a block of ground about 1,000 feet northeast of the portal of level 5 and extending for 500 feet toward the
breast the vein has not been discovered, and on levels 6 and 8 in this
block of ground its position is unknown. Presumably it has been
shifted by faults which cross the great strike fault above mentioned
and which may displace it also. To the northeast of this block of
ground the vein continues with its usual dip and strike for 1,500 feet.

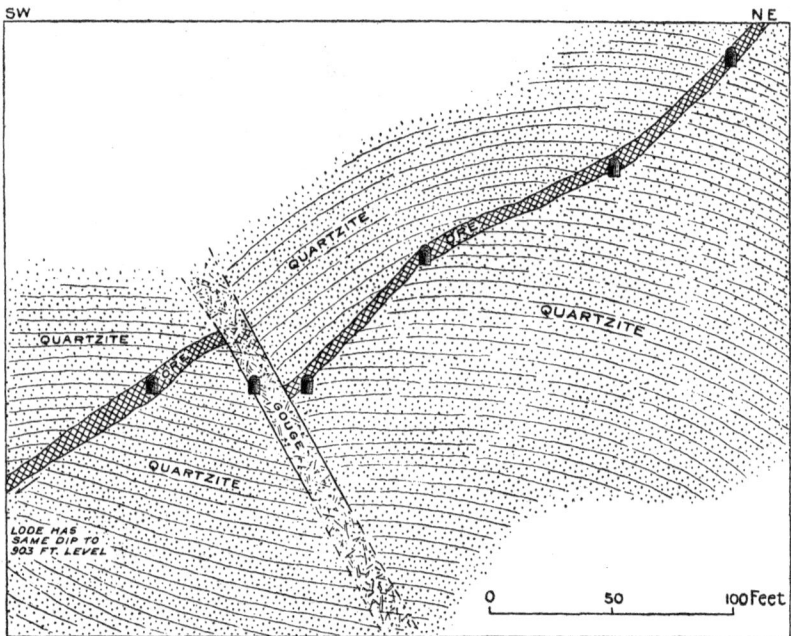

FIGURE 12.—Cross section of Lucky Boy vein, Edgemont, looking southwest, 775 feet southwest of intersection of vein with Gossip adit. Line of section trends N. 35° W.

The Lucky Girl vein, which lies several hundred feet northwest of
the Lucky Boy, strikes N. 50° E. and dips from 35° to 45° SE. On
the surface its outcrop is clearly exposed below the company bunk
house. It cuts across massive quartzite which on the foot wall dips
22° NE. An incline is driven on the vein from which two levels are
turned and a considerable portion of the lode above the lowest level
has been removed. The vein is from 2 to 7 feet wide and resembles
the oxidized portion of the Lucky Boy vein, being composed of banded,
sheeted quartz, stained yellow with iron oxide and carrying several
dollars to the ton in gold with about 1 ounce of silver. A section of
the vein is shown in figure 10.

The Big Bob vein, which is similar to the Lucky Boy and Lucky Girl veins, outcrops a few yards above the mill, where it dips about 15° SE. It is opened on five levels, the lowest of which is driven about 130 feet vertically below the outcrop. The larger part of the workings on this vein were under water in 1908.

Bull Run mine.—The Bull Run mine is located high on the south slope of Bull Run Mountain, which lies just north of the great canyon through which Bull Run Creek flows to Owyhee River. In 1902 a 10-stamp amalgamating mill and cyanide plant were built near the portal of the lowest adit, but according to report the successful operation of this mill was impossible owing to the prohibitive expense of freighting fuel and supplies up the steep hill to the mine. About $120,000 was recovered in 1902 and 1903, before the mill was shut down. The developments comprised about a mile of underground workings on the lode. The country rock is quartzite which carries thin layers of slightly micaceous shales. The quartzite, which resembles that of the Lucky Boy, is strongly sheeted and locally the

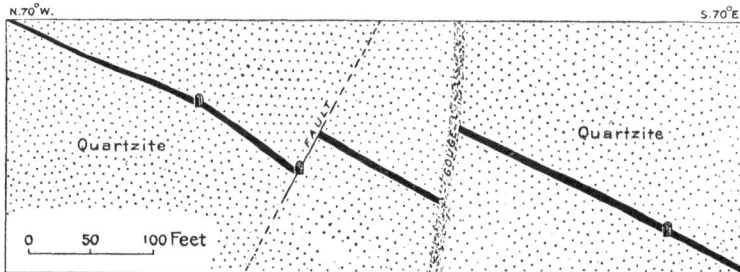

FIGURE 13.—Generalized section through Bull Run mine, S. 10° E. from tunnel 1 to tunnel 4.

shales have been contorted by compressive movements that took place before the deposition of the ore. The deposit is a well-defined fissure vein which has a maximum width of about 6 feet. It strikes northeast and dips from 22° to 38° SE.

The quartzite is sheeted parallel to the vein and in places its apparent stratification is parallel to the vein, but on the surface, where the true dip may more easily be made out, it strikes nearly eastward. The ore is composed of white quartz, stained here and there with iron oxides. In the lower levels pyrite and galena are present in small quantities, these sulphides having been leached out in the upper workings. The ore is said to carry several dollars to the ton in gold, a large proportion of which is free. The silver present is practically negligible. The composition of the ore, the country rock, and the structural features of the vein resemble very closely those of the deposits of the Lucky Girl group at Edgemont, which is not more than 2 miles to the north. The lode outcrops plainly in the steep cliffs of the mountain and may be followed through a vertical range of about 400 feet. As shown in figure 13, the lode is displaced by two faults, both of which cross tunnels 2 and 3. The faults strike about N. 15° E. and displace the lode from 30 to 60 feet. The first one encountered in tunnel 3 dips 60° W. and the second is a wide crushed zone which shows slickensided planes dipping toward both walls with numerous fragments of quartz and much clay gouge. In both faults the west wall has dropped.

Between the Lucky Girl group of veins and the Bull Run mine there is a steep canyon which is occupied by a small stream that flows westward to Bull Run Creek. A good deal of this country is covered and so the conditions for prospecting are not favorable, but its position in the line of strike of the Bull Run and Edgemont veins would seem to warrant a closer scrutiny than it has received.

MOUNTAIN CITY.

LOCATION AND HISTORY.

Mountain City is situated in the northeastern part of the Centennial Range on the north fork of the Owyhee, about $1\frac{1}{2}$ miles east of the boundary of the Duck Valley Indian Reservation, some 40 miles by stage from Tuscarora. The first discoveries were made in 1869 by Jesse Cope and others who were on their way from Silver City, Idaho, to the White Pine district, Nevada, and from this circumstance the Mountain City region is called the Cope mining district. In the seventies there was considerable activity in mining and three silver mills were in operation. These were small amalgamation mills of the Washoe pattern, equipped with stamps, pans, and settlers. It is said that over $1,000,000 in silver was recovered prior to 1881, mainly from surface and shallow workings. Since 1881 considerable prospecting has been done, but the production of ore has been small. In late years three gold mills have been built and are still in good condition, but they were not running in the summer of 1908, when the camp was visited.

GEOLOGY.

The rocks at Mountain City are limestones and shales intruded by granodiorite and overlain by rhyolite and basalt. On the summit and south slope of California Hill, 2 miles south of Mountain City, and extending northwestward from that point, is a thick bed of light-buff marbleized limestone. On the north slope of this hill, near intruding granodiorite, tremolite, actinolite, epidote, white mica, garnet, and other silicates of contact-metamorphic origin are developed in the limestone. The granodiorite is composed of plagioclase, orthoclase, quartz, mica, and hornblende. In the coarser varieties, as at the Protection mine, some of the crystals of feldspar are 1 inch long. The granodiorite is cut by aplitic dikes composed of quartz and orthoclase. Around the border of the mineralized areas there are extensive flows of rhyolite, the commonest variety of which is a purplish-pink rhyolite with phenocrysts of quartz, feldspar, and a little black mica. Other varieties of the rhyolite are flow breccias, brown glass, and black obsidian. Vesicular basalt, rich in olivine and pyroxene, was noted on the surface east of the Resurrection claims. The granodiorite is younger than the limestone, which it intrudes, causing contact metamorphism. Where the rhyolite flows were noted above the granodiorite the contact relations could not be made out, but from consideration of areas elsewhere the rhyolite and basalt are regarded as of later origin than the granodiorite, and at Mountain City they are probably later than the deposition of the silver ores.

ORE DEPOSITS.

The ore deposits are fissure veins in granite and in metamorphosed limestones. They outcrop plainly at the surface, where some of them carry good values in silver. The veins do not fall into well-defined parallel systems but strike in various directions, the prevailing dip being toward the south.

Some of the veins, as shown on California Hill, are later than the aplitic phase of the granite which cuts the normal coarse-grained granite. None of the developed deposits are in rhyolite or basalt, although some gold-bearing veins in rhyolite are said to occur in the country east of Mountain City. In the Nelson mine the lodes pass from granite to limestone without much change in width or value. There is little replacement of the limestone, for the walls are clear cut and angular fragments of the country rock are included in the veins. Where the wall rock is granite the dark silicates have been leached out and sericite and pyrite have been developed in the granite by secondary processes.

The unoxidized ore is composed of quartz, pyrite, galena, zinc blende, gray copper, argentite, gold, and arsenopyrite, with a little chalcopyrite. All of the ore is highly siliceous, quartz constituting as much as 90 per cent of the rock. The oxidized ore is composed of quartz, chalcedony, horn silver, pyromorphite, iron oxides, native gold and silver, lead carbonate, copper carbonate, and copper silicate. Brittle silver and dark ruby silver are said to be present also. The oxidation of the deposits is erratic, the sulphides occurring at some places within a few feet of the surface, while some of the minerals of oxidation are to be found as deep as the lodes have been explored, or about 250 feet below the surface. The greater proportion of the silver values are in decomposed chloride and lead carbonate ore. Specimens of rich ore show large flakes of greenish-yellow horn silver deposited in the cavities of dark quartz. Some of the iron-stained siliceous ore pans gold liberally.

The lodes are fractured and faulted, and locally the ore is reduced to a white sand, in which there are numerous small rounded fragments of quartz about the size of a hazel nut. The faults that cross the lodes are mainly of the normal type, the hanging wall having dropped with respect to the foot wall.

MINE DESCRIPTIONS.

Protection mine.—The Protection mine, located three-fourths of a mile below Mountain City, was one of the early discoveries of the district and was worked in the early seventies, when considerable chloride ore is said to have been treated in a silver mill near by. In late years the mine has been reopened and considerable exploration work has been done. A 10-stamp amalgamating and concentrating mill was built near the portal of the tunnel to treat the ore. At present part of the mine is leased and is being worked in a small way, but the mill is shut down. The principal vein is a fissure filling in granodiorite and has a maximum width of about 4 feet. The sulphide ore is composed of quartz, pyrite, galena, zinc blende, gray copper, brittle silver, and ruby silver. The surface ore is stained with iron and manganese oxides and contains horn silver, a little copper carbonate, pyromorphite, and a yellowish-green mineral, said to be silver bromide. The sorted ore carries $100 a ton in silver and gold. At some places near the vein the granite wall rock is but little altered; at others it is a light-colored decomposed rock, the ferromagnesian minerals having been leached out and the feldspar sericitized. A shaft is sunk to a depth of 62 feet and a level turned at the bottom. This is connected with an adit driven 80 feet below the bottom of the shaft, which gives a depth of 142 feet at this place.

The Protection vein strikes a few degrees west of north and has been followed into the hill on the adit level for a distance of some 750 feet to a point where it abuts against a fault that strikes eastward and dips about 40° N. A drift has been run on this fault for 400 feet, but no vein has been encountered on the hanging-wall side at this end of the drift. On the other side of the fault a vein with the same general dip and strike as the Protection vein is 35 feet farther south and is possibly the same vein, but if so the fault is reverse— a rare type of faulting in this part of Nevada. Along this fault there are stringers of quartz in place, and in the level above both the Protection vein and the faults are mineralized on both sides of their intersection. The relations indicate that the Protection fissure was displaced by faulting before deposition of the ore and that there has been considerable movement subsequently.

Resurrection mine.—At the Resurrection mine, a few rods north of Mountain City, a large amount of work has been done in tunnels, pits, and shallow inclines, but most of the workings were inaccessible when the mine was visited in 1908. The country rock is granodiorite, to the east of which are flows of rhyolite and basalt. The granodiorite, which is highly altered, is sheeted by closely spaced fissures that strike northeastward. Several narrow quartz veins cut the granodiorite parallel to the sheeting. The surface ore is composed of quartz, horn silver, lead carbonates, and iron oxide; the sulphides are galena, gray copper, a little pyrite, and chalcopyrite. In the seventies considerable rich chloride ore was taken from the surface pits and worked in silver mills near by.

Nelson mine.—The Nelson mine is on a branch of the north fork of the Owyhee, about 1¾ miles above Mountain City. Some 4,000 feet of workings have been run, mainly on two adit levels driven at a difference in elevation of about 100 feet. When the mine was visited in 1908 only the lower adit was accessible. A mill recently built at the portal of the lower adit is equipped with Blake crusher, nine stamps, amalgamation plates, and three Wilfley tables, and has treated a small amount of ore. The country rock of the mine consists of granodiorite, limestone, and aplite. The granodiorite intrudes the limestone and causes contact metamorphism with the development of epidote, actinolite, garnet, and mica in the limestone. In places this rock is so rich in actinolite that it has the appearance of a basic igneous rock and has been mistaken for diabase. The granodiorite is cut by aplite, which occurs as dikes and irregular intrusive masses. The ore deposits are fissure fillings from 1 to 3 feet wide and occur in granodiorite, limestone, and aplite. Several veins outcrop boldly on the hill above the mine, cutting across beds of metamorphosed limestone. The veins cross the contact of igneous and sedimentary rocks unbroken, but have been developed mainly in the granodiorite. The sulphide minerals present are quartz, pyrite, galena, zinc blende, gray copper, chalcopyrite, and arsenopyrite, with here and there a small amount of ruby silver and argentite. Native silver and horn silver are present near the surface, where the ore is stained with copper carbonates, iron oxides, and manganese oxides. Free gold, some of it with the crystal form, is associated with quartz and brown iron oxide. The sorted ore carries good values in both silver and gold, some specimens containing a high percentage of horn silver.

The Standard vein, which is developed in the lower tunnel, strikes southeastward and has been followed for about 1,000 feet, with overhead stoping here and there. This vein is faulted at three places by faults that strike eastward and dip northward at various angles. One of the faults shows a horizontal displacement of about 150 feet, the other two of less than 15 feet. All are of normal type, the hanging wall having dropped with respect to the foot wall.

Mountain City mine.—The Mountain City mine is located about 1 mile southwest of Mountain City, at the top of a low, flat ridge that rises some 200 feet above Owyhee River. The country rock is a metamorphosed black, shaly limestone which strikes eastward and dips 50° N. The lode is a fissure vein which cuts across the limestone, striking N. 50° W. The ore is highly siliceous and is a simple fissure filling, cementing angular fragments of the altered limestone. It carries silver chloride and native silver, and in the seventies, according to report, several hundred thousand dollars' worth of silver ore was taken from the deposit through a shaft now inaccessible. About 500 feet S. 75° E. of the principal workings of the old Mountain City vein and lower on the hill are a number of open pits, some of which have been sunk on a vein which strikes N. 32° W. Possibly it is the faulted continuation of the Mountain City vein, but this has not yet been determined. This ground has recently been acquired by J. Hall and others, of Mountain City, and is called the New Yorkeys claim. The country rock of the lower deposit is a dark-gray metamorphosed limestone flaked with tremolite crystals. A tunnel is driven 95 feet N. 70° W. to the vein, which it follows for 90 feet, and a winze is sunk on the ore body 60 feet below the adit level. The deposit is a fissure vein and at some places a sheeted zone composed of several narrow veins with slabs of limestone between. Much movement has occurred since deposition, for at places the quartz is brecciated almost to powder. The ore is composed of quartz, iron oxide, copper carbonates, and silicates, and a little pyrite is present at the bottom of the winze. The vein strikes N. 32° W. and dips from 56° to 85° S. It has a maximum width of 5 feet, and is said to carry good milling values.

VAN DUZER CREEK PLACERS.

Van Duzer Creek is a small stream which flows eastward, joining the north fork of the Owyhee some 6 miles south of Mountain City. About 2 miles above the point where the stage road crosses the stream and extending westward up the main fork for about a mile portions of the channel have been washed for placer gold. Two strips of the channel, from 20 to 60 feet wide, have been worked out, one of these for a distance of 1,200 feet and the other for about 1,000 feet. Two hydraulic plants with 10-inch pipe and monitors are installed along the stream about three-fourths of a mile apart. The depth of work is nearly everywhere less than 15 feet. The ground is mainly fine gravel with few small bowlders, and the bed rock is presumably limestone. The gold is said to vary from fine dust to nuggets of 5 or 6 ounces and sells for $17 an ounce. The mines, which were discovered by R. M. Woodward in 1893, have been worked in some seasons for about fifteen years, but in 1908 were idle. The source of the gold is presumably some undeveloped veins at the head of the stream.

LONE MOUNTAIN.

Lone Mountain, which is called Nannies Peak in the reports of the King Survey, is a striking topographic feature rising conspicuously above the main axis of the Seetoya Range to an elevation of 9,046 feet. The mountain mass is composed in the main of Carboniferous limestone—dark-blue or gray massive beds locally metamorphosed by intruding igneous rock. The prevailing dip of the limestone is from 30° to 80° W. The crest of the mountain is composed of quartz monzonite and quartz monzonite porphyry, which cut through the limestone and inclose small blocks of it. On the west slope of the mountain, from 200 to 700 feet below the summit, is a long, narrow intrusive mass of felsitic quartz porphyry which trends nearly due north. Certain phases of the porphyry are almost as dense as rhyolite; others resemble the common types of quartz porphyry. The quartz monzonite and the porphyry locally grade one into the other and are probably phases of the same intrusion. The quartz monzonite, which is of medium grain, is composed of feldspar, quartz, hornblende, and biotite. An analysis of this rock is given on page 26, where the granitic rocks are described. The porphyry has a finely crystalline groundmass composed of quartz and orthoclase, in which are embedded phenocrysts of acidic plagioclase and resorbed quartz, biotite, and hornblende.

Both the quartz monzonite and the quartz monzonite porphyry have caused contact metamorphism of the limestone. At some places the limestone is changed to a garnet-calcite rock, at others to a green actinolite rock, at others to a hard cherty hornstone, and at still others to a coarse-grained marble.

The Merrimac district is situated on Lone Mountain about 28 miles by wagon road northwest of Elko. More or less prospecting has been done in this district since 1879, and from various claims about 1,000 tons of ore has been shipped, with a total value of over $30,000.

The ore bodies, so far as observed, are deposits of lead and copper ore in marbleized limestone and contact-metamorphic deposits of copper ore in garnetized limestone. When the camp was visited by the writer nearly all the workings were inaccessible. The lead and copper ores in marbleized limestone are near the intruding igneous rocks and are usually highly shattered and oxidized. Some of them are at the contact of quartz monzonite or porphyry with limestone, and some of them are several hundred feet from it. The common minerals are galena, pyrite, chalcopyrite, and their oxidation products. Most of the ore consists of iron-stained limestone and quartz containing lead carbonate and green and blue copper carbonates, with copper oxides and yellow pyromorphite. Such ore is exposed at several places at the south end of the mountain on the Baltimore group of claims, and also on the Floradora and Ajax claims, a mile or more northeast of the summit. This ore is said to carry high values in silver, with $1 to $2 in gold to the ton. A number of shafts have been sunk and short tunnels have been driven on the deposits, but the present accessible workings do not show the extent or shape of the ore bodies. Some promising lead-silver deposits in limestone are said to be located on the north end of the mountain just outside of the area shown in the outline map (fig. 2).

The contact-metamorphic deposits occur in the garnet zones near the intrusive igneous rocks. The large dumps from workings now inaccessible on the Baltimore and Cuag claims show a wide zone of decomposed limestone, with some garnet rock stained with iron oxides and copper carbonates. A low-level tunnel was driven northeastward for 850 feet to explore the ore bodies in depth. It crosscuts westward-dipping limestone for 750 feet, to a point where it encounters quartz monzonite porphyry similar to that which forms a large part of the mountain crest. Near the contact the limestone is metamorphosed to garnet, actinolite, and other silicates, with which are intergrown iron and copper sulphides. Pyrite, arsenopyrite, and chalcopyrite are present, and locally copper carbonate.

A zone of metamorphosed limestone is situated on the east slope of the mountain and trends northward approximately parallel to the crest, and there are a number of pits and short tunnels which are driven in ore. On the Morgan claim and north of it for several hundred feet, the porphyry lies to the east and metamorphosed limestone to the west of the contact. Highly oxidized copper and lead ore, some of it associated with garnet gangue, is exposed at several places. On the Pacific claim, which lies to the south of the Morgan, there is a good-sized body of iron ore at the contact of limestone and porphyry. This ore is composed of magnetite, limonite, garnet, and copper carbonate. Some of the limonite is pseudomorphous after pyrite, indicating that the original deposit was composed of intergrown magnetite and pyrite. This ore is said to carry a small percentage of copper, and if it were accessible to smelters it would make a good fluxing rock.

CORTEZ RANGE BETWEEN THE CARLIN AND DALTON PEAKS.

GENERAL FEATURES.

The Cortez Range, north of Humboldt River, is a lofty ridge about 45 miles long and from 5 to 20 miles wide. The highest summits are the Carlin Peaks, which reach an elevation above 7,700 feet, and the Dalton Peaks, about 25 miles farther north, whose highest point is 9,232 feet above the sea. Rye Meadows, a broad area of gently rolling sagebrush plain, lie to the west of this portion of the range, and on the east side is the broad, gently sloping valley of Maggie Creek, trending southward approximately parallel to the range. The Carlin Peaks[a] are composed in the main of extensive flows of rhyolite, which on the west slope of the highest summit surround a large mass

---

a Emmons, S. F., U. S. Geol. Expl. 40th Par., vol. 2, 1877, p. 587.

of andesite. The rhyolite series is probably not so thick as would appear from the extent of the flow and from the differences in the elevations at which it is found, for here and there small areas of sedimentary rocks outcrop where the rhyolite has been eroded. These may be observed in the Richmond district and on either side of Maggie Canyon, where the stream cuts through the low ridge that joins Carlin with Maggie Peak. Granodiorite and other coarse-grained rocks which are probably intrusive in limestone were noted in the Richmond district on the east slope of the range, 8 miles north of the Carlin Peaks. The north end of the range, between the Lynn district and Soldier Gap and including the main mass of the Dalton Peaks,[a] is made up of the Weber quartzite, which dips toward the west and is bordered on the east by rhyolite. In this area there are several mines and prospects, none of which are extensively developed or have produced more than a few tons of ore.

MINE DESCRIPTIONS.

Nevada Star mine.—The Nevada Star mine is located 9 miles northwest of the town of Carlin, on the ridge just west of Maggie Canyon. The deposit is a small replacement vein in limestone, dips steeply eastward, and contains bunches of galena and lead and copper carbonates said to carry good values in silver. A shaft is sunk on the lode to a depth of 60 feet, and at the bottom a drift is driven southward for 80 feet. Near by a two-compartment shaft has been sunk to a depth of 110 feet, but from it no crosscutting has been done.

Copper King claim.—At the Copper King claim, which is about 2 miles southwest of the Nevada Star, the country rock is banded, fissile rhyolite, which dips about 40° N. The rhyolite is shattered, leached, and cut by slickensided movement planes that carry bunches of oxidized copper ore here and there. A well-equipped two-compartment shaft has been sunk to a depth of 150 feet, with short levels turned at intervals of 50 feet.

Richmond district.—There are several small prospects of silver ore in the Richmond district, which is on the east slope of the Cortez Range, 14 miles northwest of Carlin. The rocks exposed in this district are limestones and quartzites intruded by granodiorite. In the limestones near the intruding rock there are several small outcrops of siliceous silver ore which contains a small proportion of galena, gray copper, and lead and copper carbonates. Shallow pits are sunk on several of these outcrops.

Lynn district.—The Lynn district, just east of a low divide at the crest of the Cortez Range, is about 20 miles northwest of Carlin. Gold was discovered in this district in April, 1907, and during the following summer the camp was the scene of one of the rushes which

_____a Emmons, S. F., U. S. Geol. Expl. 40th Par., vol. 2, 1877, p. 608.

often follow discoveries in Nevada. The excitement had subsided in 1908, and the district was deserted by all except about a dozen miners who remained to develop their claims.

The country rock is bedded rhyolite, and in the vicinity of the principal deposits it dips westward at low angles. On the west side of the divide, a few hundred yards away, an outcrop of decomposed andesite was noted. At the Big Six mine the bedded rhyolite strikes S. 30° W. and dips from 20° to 35° W. Along a zone of shattering the rhyolite is silicified, is stained with iron oxide, and carries bunches of rusty ore which pans gold freely. Three inclines, 20, 40, and 65 feet long, are driven down the bedding planes of the rhyolite, and a few tons of ore has been shipped from these workings.

At the Gold Dollar mine, about 200 yards N. 10° E. of the Big Six inclines, a zone of shattering strikes northwestward across the rhyolite, and along this zone veinlets of quartz and silicified iron-stained rhyolite carry high values in gold. Four deep trenches are dug across the zone of shattered rhyolite.

Several placer claims are located in the gulches which drain eastward from the Big Six and Gold Dollar mines, and at a number of places the gravels carry placer gold. On the Hilltop claim, belonging to Hugh Jones, a block of gravel 60 feet long, 8 feet wide, and about 5 feet deep yielded $150. During the summer of 1907 about $1,000 was taken from this gulch. In the upper part of the gulch the bed rock is rhyolite, but this has been cut away lower down, locally exposing the underlying limestone. It is thought that water to be used for sluicing could be collected on the slopes of the Dalton Peaks, if the deposits prove extensive enough to warrant it.

PINYON RANGE.

GENERAL FEATURES.

The Pinyon Range is a long, narrow group of hills and mountains which extends southward from Humboldt River to a point near Eureka. Its highest point, called Ravens Nest, is 8,386 feet above sea level, or more than 3,000 feet above Pine Valley, which lies to the west of the range.

The rocks that constitute the range are in the main Paleozoic sedimentary rocks, including the section from the Ordovician to the Carboniferous. In the vicinity of the Bullion mining district these sedimentary rocks are intruded by granodiorite. At the north end of the range and at many places along its flanks the sedimentary rocks are covered over by rhyolite, basalt, and the Humboldt formation.

As shown by the analytical map of the Fortieth Parallel Survey (Pl. XI, vol. 1), the structure of the range is anticlinal from Ravens Nest to Pinyon Pass. From Mineral Hill to Diamond Valley the rocks are compressed into an open syncline, the axis of which trends a few degrees west of north. South of this point the structure is that of an eastward-dipping monocline. In connection with the studies at Eureka, Nev., C. D. Walcott made a section across Pinto Peak for the purpose of comparison with the section at Eureka.[a] This section (fig. 14) is given herewith and shows that the anticlinal structure of the range is modified to an important extent by faulting.

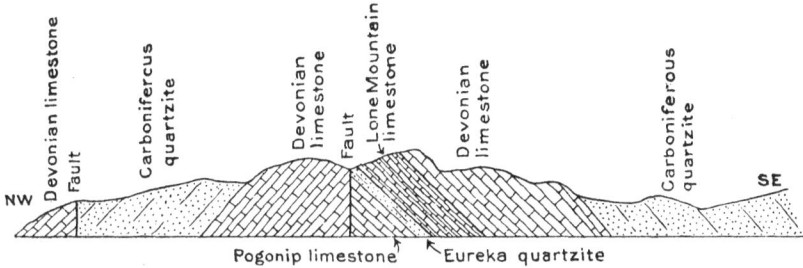

FIGURE 14.—Cross section through Ravens Nest, Pinyon Range. After C. D. Walcott.

Northwest of Pinto Peak, just above the valley flat, is a dark-blue limestone which carries Devonian fossils. This rock is faulted against Carboniferous quartzite that contains conglomerates and black siliceous pebbles. These dip northwestward and are underlain conformably by blue limestone which carries Upper Devonian fossils. High on the mountain, northwest of its summit, the limestones are faulted near the crest of the anticline. To the northwest of this fault the beds dip westward, but southeast of it they dip east. East of the fault is dark ferruginous Eureka quartzite, overlain by light-gray siliceous limestone, which contains *Halysites* and is probably of Ordovician age. The beds overlying the Ordovician carry an early Devonian fauna and above these Upper Devonian species. Above the Devonian is a great thickness of quartzites and sandstones with some argillaceous beds. The Upper Devonian and the Carboniferous dip away from the fault at the crest of the anticline on either side.

BULLION.

GEOLOGY AND ORE DEPOSITS.

Bullion, which is the principal settlement in the Railroad mining district, is about 28 miles southwest of Elko and 12 miles southeast of Palisade. It is situated at the base of the east slope of the Pinyon Range near the headwaters of the west fork of Dixie Creek, at an elevation of about 6,600 feet. The mines were discovered in the late sixties and were successfully worked in the seventies and eighties, when two small smelters were in operation. In 1906 some of the mines were reopened and since that time a moderate tonnage of ore has been taken from the old workings, together with several carloads

---

[a] Hague, Arnold, Geology of the Eureka district, Nevada: Mon. U. S. Geol. Survey, vol. 20, 1892, p. 201.

of slag from the slag dump of the lead smelter at Bullion. The district is said to have produced about $3,000,000 in silver, lead, copper, and gold. At present the principal holdings are owned by or are under option to three companies, each of which is actively engaged in mining or in exploration work. They are the Nevada Bunker Hill Mining Company, the Trimetal Mining Company, and the Delmas Copper Company. Two of these companies are driving long low-level tunnels below the old workings, giving vertical depths up to 1,500 feet below the surface. From these tunnels it is planned to prospect a considerable area of promising ground. At present the ore from the Standing Elk and near-by mines is hauled by wagon 28 miles to Elko. From the Delmas property on the west side of the range ore is hauled 12 miles to the Eureka and Palisade Railroad, to be reloaded at Palisade into broad-gage cars.

The mines are situated high on the slopes of Bullion Hill, a spur of Ravens Nest. This mountain is composed in the main of Ordovician limestone, which is for the most part a gray marbleized limestone with a general westward dip. On the northeast slope of the mountain there is considerable faulting and the abrupt changes of dip indicate a complex geologic structure. The limestone is intruded by granodiorite and by quartz porphyry.

The granodiorite, which is of medium grain, is composed of quartz, feldspar, and biotite. Under the microscope the feldspars are seen to be oligoclase and andesine, with only a little orthoclase. The rock approaches quartz diorite in composition and is poorer in potash than most granodiorites. It forms extensive outcrops near the summit of the range and causes considerable contact metamorphism of the Ordovician limestone which it intrudes. Near the granodiorite the limestone is changed to a rock composed of garnet, tremolite, actinolite, calcite, andalusite, and white mica, which at some places are intergrown with pyrite, chalcopyrite, and zinc blende.

The intruding quartz porphyry is a light-colored rock which contains abundant rounded phenocrysts of quartz and a smaller number of feldspar phenocrysts, in a light-colored, fine-grained groundmass. Muscovite and some biotite are present, but the dark-colored silicates are very sparingly developed and are much less abundant than in the granodiorite. The quartz porphyry forms large intrusive masses on the northeast slope of Bunker Hill and is exposed at several places in the Standing Elk and Tripoli mines, where it is greatly altered by hot waters, white mica and carbonates being extensively developed in the groundmass. The limestone near the quartz-porphyry intrusions is somewhat marbleized, but garnet zones are not developed along its contacts.

The most important ore bodies are replacement deposits of lead, silver, and copper ore in marbleized limestone, and copper deposits of contact-metamorphic origin. At some places the quartz porphyry is impregnated with copper minerals and possibly some of this rock could be worked under favorable conditions. There are also auriferous quartz veins in the granodiorite, but these, so far as developed, are of small economic importance. The replacement deposits of lead, silver, and copper ore in marbleized limestone are commercially of greatest importance, and most of the production from the district has come from ore bodies of this character. Most of these deposits were inaccessible when the mines were visited, but those which were exposed appear to be nearly vertical chimneys of ore situated at the intersection of two or more relatively narrow replacement veins. The maximum depth to which the ore is exposed is about 500 feet below the surface and the ore is almost completely oxidized. The principal ore minerals are lead carbonate, horn silver, pyromorphite, malachite, azurite, chrysocholla, cuprite, pyrite, chalcopyrite, galena, bornite, copper glance, and a copper-antimony mineral that is probably gray copper. The gangue is composed of quartz and calcite. The ore is nearly everywhere stained by red and brown iron oxides, which, together with calcite and quartz, form the most abundant minerals. A little manganese oxide is present.

The deposits outcrop as iron-stained gossans where the surface is strewn with craggy bowlders composed of iron oxides, with copper carbonates here and there. Some of the gossans have not been explored in depth, and none of the deposits have been followed down to pure sulphide ores. Owing to the relief of the country and the shattered condition of the rocks, oxidation has extended to considerable depth.

The replacement deposits of copper ore of contact-metamorphic origin have received serious attention only within the last two years, probably because the mass of this ore is of too low grade to stand excessive freight and treatment charges. The Delmas Copper Company has lately sunk a number of shallow pits and has driven several short tunnels on these deposits, exposing a considerable quantity of this ore, some of which has been shipped to Utah smelters.

On the summit of the ridge which divides the drainage of Pine Creek from that of Dixie Creek the intruding granite porphyry is in contact with the Ordovician limestone. The line of contact stretches northeastward and may be followed for nearly 2,000 feet down the southwest slope of the ridge and for a considerable distance down the northeast slope. The zone of limestone along the margin for a width of 125 to 400 feet has been converted into a rock composed of garnet, actinolite, calcite, epidote, quartz, tremolite, zoisite, and pyroxene, and at many places copper-bearing sulphide and zinc blende are intergrown with the iron-bearing silicates. The principal ore minerals are pyrite, chalcopyrite, bornite, chalcocite, galena, and zinc blende. These are locally altered to malachite, azurite, copper oxides, iron oxides, and chrysocolla. The ore of the better grade is said to carry from 3 to 8 per cent of copper with good values in silver. Although much of the garnet rock is barren of copper minerals, the work that has been done seems to show that some of the ore bodies are of considerable extent. The boundaries of the ore zone are very well defined. The contact between the garnet rock and the marble which cuts across the bedding is usually rather sharp and clean cut. Near the top of the ridge a vein of garnet rock about 1 foot wide cuts across the bedding of the marble, and just below this vein

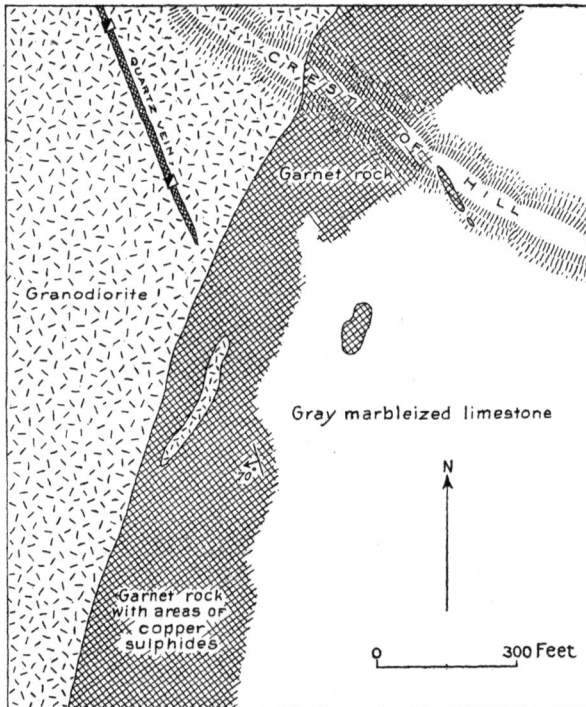

FIGURE 15.—Sketch showing the relation of the ore zone to granodiorite and limestone on the Sweepstakes claim of the Delmas group, Bullion district.

small patches of the garnet rock are surrounded by the marble. The garnet, in its relations to the marble, resembles some deposits of siliceous ore in limestone and indicates that in this particular locality the garnetization follows fissures in the rocks. In the main, however, the garnet rock follows the contact of the limestone with the granodiorite, and no garnet veins were noted more than a few hundred feet away from the igneous rock. Figure 15 is a sketch showing the relations of granodiorite and garnet rock on the Delmas claims.

The deposits of quartz porphyry impregnated with copper minerals which have been developed through exploration for lead-silver ores are too low in grade to work under present conditions. In the Standing Elk mine, in tunnel 5, the quartz porphyry intrudes the limestone. It contains much hydrous silica and is, for a distance of about 20 feet, extensively altered by hot solutions that have replaced the feldspars and the groundmass with sericite or white mica, which is now stained with copper carbonate. The feldspars have been leached out and their spaces are now filled with fibrous malachite and with chalcanthite. The rock is said to carry 2 per cent of copper and a small amount of silver.

The auriferous quartz veins in granite porphyry are, so far as developed, small and of too low grade to work. On the Delmas ground one of these is exposed here and there for a distance of some 500 feet along the strike. It is from 2 to 4 feet wide, is composed of white quartz, limonite, and pyrite, and is said to carry about $3 to the ton in gold. Several smaller veins of this character outcrop in the areas of granodiorite between the Delmas claims and the Copper Belle, but none of them have been extensively explored.

MINE DESCRIPTIONS.

Standing Elk mine.—The Nevada Bunker Hill Mining Company controls the Standing Elk, Tripoli, Red Bird, and other mines, and is driving a crosscut tunnel to intersect the several lodes in depth. This tunnel in July, 1908, was 1,500 feet long but was still several hundred feet from the nearest lode. The Standing Elk, the most important mine of this group, is opened on seven levels, mainly adits, having altogether a vertical range of 600 feet. There are several thousand feet of workings on this claim, the principal level being adit No. 5. When the camp was visited the workings above this level were not accessible.

The ore bodies are irregular replacement veins in limestone, which intersect to form chimneys of ore, the largest being about 50 feet in diameter. The country rock varies from a hard gray limestone to a massive marble which locally is very coarsely crystalline. The limestone is sheeted, brecciated, and filled with white calcite. In a few places garnet rock is developed in the mine, but none of the Standing Elk deposits consist of the typical garnet ore. The intruding quartz porphyry is much decomposed and is locally silicified, and a considerable mass of it carries copper. Nearly everywhere the ore is highly oxidized. The principal ore minerals are lead and copper carbonates, copper and iron oxides, bornite, pyrite, chalcopyrite, and a copper-antimony sulphide, which is probably gray copper. Calcite and quartz are the important gangue minerals; a little fluorite is present in microscopic crystals, intergrown with quartz.

Tripoli mine.—The Tripoli mine, about 800 feet northeast of the Standing Elk, is owned also by the Nevada Bunker Hill Mining Company. A tunnel is driven into the mountain for 100 feet to an engine station, where a winze is sunk to a depth of 175 feet. Levels are turned at vertical intervals of about 50 feet and a second winze is sunk from the lowest level, giving a maximum depth of 300 feet below the surface. The country rock is marbleized gray limestone which on the surface strikes N. 20° W. and dips from 85° SW. to 90°. The limestone is intruded by an acidic phase of granodiorite. The incline and drift are driven on a fissured zone, which strikes about N. 25° W. and is approximately vertical. The zone of crushed limestone is crossed by two fissures, approximately at right angles, and these dip steeply toward the northwest. Chimneys of rich silver-lead ore are formed at or near the intersections. Lead carbonate and pyromorphite are mixed with galena and copper carbonates, and the ore is highly oxidized as deep as development goes. In the bottom of the winze, 300 feet below the surface, garnet and tremolite are extensively developed. At this place the metamorphosed rock is crushed and greatly altered, and it is said to carry about $25 in lead and silver.

At the surface to the east of the Tripoli claim the limestone is medium-grained marble containing reefs and patches of garnet rock, with which are intergrown small masses of pyrite and chalcopyrite, stained here and there with copper carbonates.

Red Bird mine.—The Red Bird mine, of the Bunker Hill group, is about half a mile northeast of the Standing Elk and nearest the portal of the low-level adit. A tunnel is driven for 250 feet on a steeply dipping lode, and 80 feet below this a second tunnel is driven for 200 feet. About 150 feet from the portal of the lower tunnel a raise connects the two levels. The lode, which at some places is 4 feet wide, carries good values in lead and silver.

Copper Belle mine.—The Trimetal Mining Company controls a large acreage which joins the Nevada Bunker Hill holdings on the northwest. This group includes the Copper Belle, Copper King, and Philippine claims, on each of which considerable development work has been done. The Copper Belle mine is about half a mile northeast of the Standing Elk. The principal tunnel is driven for 325 feet S. 60° E. to the ore body, which is a large irregular mass of oxidized ore carrying lead, silver, and copper. It is said to have produced about half a million dollars' worth of ore, which was smelted in the copper smelter at Bullion. The ore body, which is nearly everywhere inaccessible, resembles the deposits of the Standing Elk in that it is a replacement of marbleized limestone. On the surface above the ore body and at several other places near by the rock is stained with iron oxides and copper carbonates.

Delmas mine.—The Delmas Copper Company owns several claims which are located near the crest of the mountain range and extend southwestward from it.

On the Sweepstakes claim, which is the most extensively explored, are the copper deposits of contact-metamorphic origin which have been mentioned above. These deposits, as already stated, are composed of garnet, tremolite, and other contact-metamorphic minerals, which at some places are intergrown with pyrite, chalcopyrite, bornite, galena, and zinc blende. Locally the copper-iron sulphides, which are unquestionably primary, are coated over with secondary copper glance. Here and there the copper sulphides have oxidized to copper carbonates and iron oxides, but the oxidation is not complete, even at the surface, and at some places less than 10 feet below the surface the sulphides are much more abundant than the carbonates and oxides. Shipments of 210 tons of ore from this claim averaged 70 ounces of silver to the ton, 10.4 per cent of copper, and 2.8 per cent of lead. There is a considerable tonnage of low-grade ore partly developed. It will probably be found that the high-grade ore is restricted to that which carries chalcocite-coated sulphides, or to that which has resulted from the oxidation of such ore. Figure 15 is a sketch of the Delmas property.

Other claims.—The Kenilworth claim, north of the Standing Elk; the Sylvania claim, west of the Standing Elk; and the Blue Belle, northwest of the Sylvania, have each produced considerable ore from workings which were inaccessible when the camp was visited by the writer. The principal deposits of these mines appear to resemble those of the Standing Elk rather than the contact-metamorphic deposits of the Delmas group.

MINERAL HILL.

LOCATION AND HISTORY.

Mineral Hill is a mining camp situated about 5 miles southeast of Mineral, a station on the Eureka and Palisade Railroad. It is at the north end of a small ridge of the same name which rises some 700 feet above the floor of Pine Valley and forms a foothill on the west slope of the Pinyon Range.

The deposits, which outcrop along the summit of Mineral Hill, were discovered by a party of prospectors from the Reese River district in 1868 and were sold soon afterward to a San Francisco company. In the early seventies the mines were sold again to a London corporation, which operated them with some success until 1878. Parker, Spencer & Co., who were mining at the south end of the hill, acquired the holdings of the London company in 1880 and operated the mines and mill until 1887. Since that time the mines have not been worked actively. The total production of Mineral Hill, so far as it can be estimated from various reports, is probably a little more than $6,000,000, practically all of which is silver. The mines are now in the hands of the Mineral Hill Consolidated Mining

Company, with headquarters in New York, and this company plans to work the low-grade ore that remains in the dumps and mines.

The ore was treated in two silver mills, one of which was equipped with 15 and the other with 20 dry-crushing stamps. The 15-stamp mill was in operation for a number of years, but the larger mill was sold and removed after it had been running for only a short period. The mills were equipped with Stetefeldt roasters to work the ore by the Reese River process, but it was found that the additional cost of roasting was greater than the increased saving effected, and so the Washoe process with dry crushing and raw amalgamation was early adopted. The details of treatment are given by M. Eissler in the "Metallurgy of silver," page 154. The present owners plan to treat the ore by concentration and cyanidation.

GEOLOGY.

The Pinyon Range to the east of Mineral Hill is made up of steeply dipping sedimentary rocks consisting of Paleozoic limestones, quartzite, and shales. The ore deposits are in a gray crystalline limestone, which, along the crest of Mineral Hill, dips from 45° to

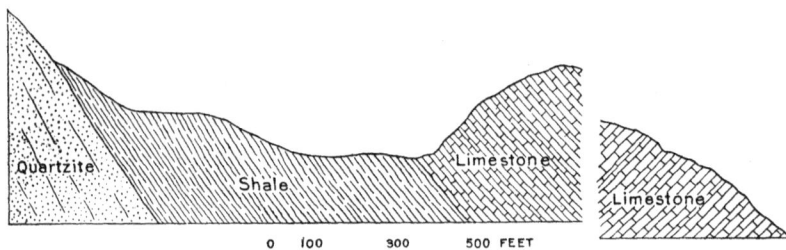

FIGURE 16.—Sketch illustrating the structure at Mineral Hill.

75° E. This limestone rests upon a dark shale that is exposed along the west slope of the hill and in the Taylor tunnel below the ore bodies. On the west slope of the Pinyon Range, east of the south end of Mineral Hill, the shales, which here are at least 600 feet thick, dip westward below the limestone and rest upon a great series of quartzites which are mapped as Ogden (Devonian) in the atlas of the Fortieth Parallel Survey. Figure 16 is a sketch in cross section drawn westward through Mineral Hill.

The structure appears to be synclinal, the axis of the syncline lying somewhere to the east of the crest of Mineral Hill and east of the outcrops of the ore deposits. The limestone that contains the ore bodies is about 400 feet thick, but it is probably only the lower part of a formation which may be thicker, the upper portion having been eroded away at this place. The limestone is cut by three narrow dikes of a decomposed intruding igneous rock, the least-altered specimens of which are composed of quartz, sericite, calcite, and limonite. These dikes are approximately parallel, strike eastward, and dip to the south at high angles.

ORE DEPOSITS.

The principal workings of Mineral Hill are open cuts and shallow stopes which are closely spaced along a zone from 200 to 300 feet wide and about 1,200 feet long. The open cuts are from 25 to 75 feet long and their width is somewhat less. The stopes range from 10 to 40 feet in width and do not extend downward more than 150 feet below the surface. The Queen tunnel is driven southward from the north end of the workings, exploring a large part of the ore zone, and ore chutes are raised to the ore bodies near the surface. About

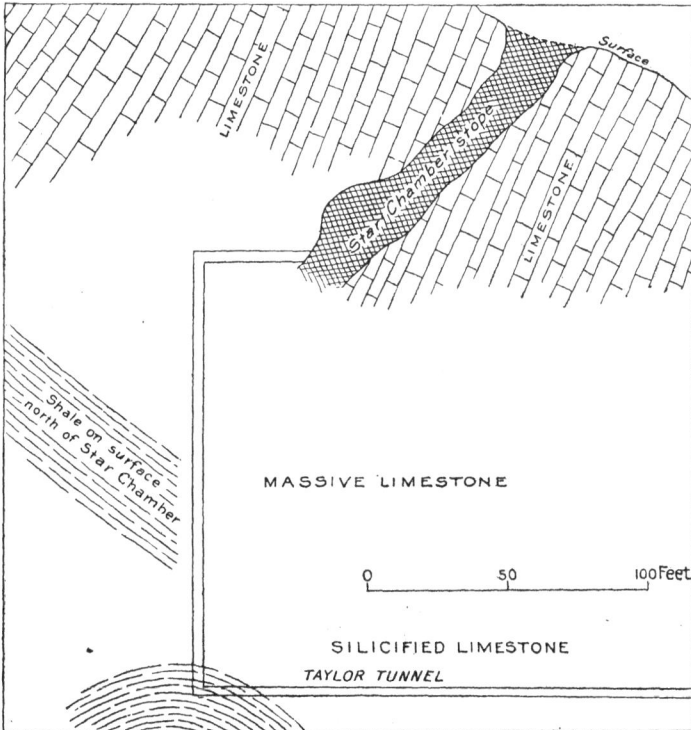

FIGURE 17.—Cross section of Star Chamber stope, Mineral Hill.

150 feet below the Queen adit the Taylor tunnel is driven westward to intersect a winze from the Queen.

The deposits are very siliceous and are eroded less rapidly than the country rock; consequently they form bold outcrops, and the zone along which a maximum of mineralization has taken place has remained longer at the surface, the hard rock tending to monopolize the outcrop. The configuration of the hill is probably due to the resistance offered to erosion by the large masses of quartz. The ore bodies are chambers or irregular replacement deposits which cut

11444—Bull. 408—10——7

across the bedding of limestone. Along the line of the principal deposits the limestone dips steeply eastward, as shown in figure 17. In places near the ore the limestone is strongly sheeted or brecciated, the open spaces being filled with white calcite, which contrasts strikingly with the gray limestone bands of the original rock, producing gneisslike banding. At some places the limestone is replaced by the quartz and sulphides and the ore grades into the country rock; elsewhere the limestone is brecciated, and the small angular fragments of limestone, which are cemented by white quartz, show no evidence of having been dissolved, the sharp edges remaining intact. The relations indicate that the composition of the solutions changed while the deposits were being formed—that they had spent their power to replace the limestone before deposition ceased. The ore, deposited by solutions which were capable of replacing the limestone, is richer in the sulphides and in silver than the quartzose material that simply filled the open spaces.

The minerals of the ore are quartz, calcite, barite, silver chloride, argentite, gray copper, galena, zinc blende, copper carbonates, pyromorphite, lead carbonate, pyrite, and iron and manganese oxides. According to Eissler,[a] polybasite, stephanite, bromide of silver, and molybdenite are also present. Some of the ore carries a considerable quantity of galena, but in most of it the proportion of sulphides is small, the gangue minerals constituting considerably more than 90 per cent of the bulk of the rock. A large proportion of the ore carried from 100 to 200 ounces of silver to the ton, although ore which carried as low as 25 ounces was worked. The ore bodies are closely spaced along a zone of fracturing that strikes northward and has been extensively explored for a distance of about 1,500 feet. So far as developed, the ore is in the main at the surface or less than 100 feet deep. Most of the ore bodies dip about 45° E.; some are vertical and one dips steeply westward. The silicified zone is cut by a number of fissures which strike east and dip about 60° N., and in two of the ore chambers these fissures form the south wall of the ore body. Possibly these fissures have faulted the ore, but the limestone has been silicified on both sides of them and the ore zone is not displaced by them to any great extent.

From the Star Chamber stope was removed a wide mass of ore, extending from the surface to the level of the main adit with a dip of about 40°. This stope, which is shown in figure 17, is practically continuous toward the south with the Giant stope, a huge cavity about 50 feet in diameter from which small tortuous stopes extend in several directions. Still farther south along the zone of silicification are the Live Yankee, the Austin, and several other stopes of smaller size. The localization of the ore bodies is due to the intensity

---

of fracturing and sheeting in the ore zone, the larger deposits having formed where there was a maximum amount of shattering.

The shales which underlie the limestone are crumpled, fissured, and cut by small veins of white quartz, but are not known to carry deposits of economic value. Below the ore bodies some development work has been done in the shales, but this prospecting was not thorough and the present owners plan further exploration at the lower level. Work to the east of that already done in the Taylor tunnel may prove more productive. To judge from the dip of the beds the limestone will probably be found at greater depth at that place than in the ground directly below the outcropping deposit.

ALPHA.

Alpha is a small camp about 15 miles south of Mineral Hill and 5 miles east of Alpha station on the Eureka and Palisade Railroad. The principal claims are the Arizona, Utah, Oregon, and Idaho. These claims have been developed by a number of shallow inclines and short tunnels, driven for the most part in silicified limestone which carries a considerable amount of barite and a small amount of the metalliferous sulphides. A concentration plant equipped with ten stamps and five vanners was built at Chimney station, about 3 miles west of the mines, but the treatment was presumably unsatisfactory, as only a small amount of ore was put through the plant. The country rock is Devonian limestone. It dips from 30° to 40° E. and presumably rests upon the quartzite which outcrops as a marginal band in the low hills west of the mines and is mapped as Ogden by the Fortieth Parallel Survey. The ore deposits outcrop boldly at the surface, some of them forming more or less noticeable reefs. The metalliferous minerals include freibergite, galena, zinc blende, pyrite, and copper carbonates, and in some places barite forms more than half of the ore. The lodes are sheeted zones and replacement deposits in limestone, of which some follow the stratification and some cut across the bedding.

CORTEZ RANGE SOUTH OF HUMBOLDT RIVER.

GENERAL FEATURES.

The Cortez Range extends from Carico Peak northeastward about 50 miles to Humboldt River and from this point northward some 40 miles to Independence Valley in the region of Tuscarora. The range north of Humboldt River has been described. That portion which lies south of the Humboldt includes a part of the Safford district, near Palisade, and the Mill Canyon and Cortez district, on the slopes of Tenabo Peak. The country between Mill Canyon and the Safford district was not traversed in this reconnaissance, and the notes referring to that portion of the range are taken from the descriptions given in the report of the Fortieth Parallel Survey.[a]

The south end of the range as outlined above is 50 miles long and about 12 miles wide. The highest summits are Railroad, Tenabo, Cortez, and Papoose peaks. Of these Tenabo Peak is the most elevated, its summit being 9,240 feet above sea level, or about 4,000 feet above the valley flats. The range is crossed by several low passes, among them Wagon Canyon, Agate Pass, and Cortez Pass. Crescent Valley, a great gravel-covered sagebrush plain, borders the range to the northwest; Pine Valley and Garden Valley lie to the southeast.

GEOLOGIC FEATURES.

The rocks of the Cortez Range south of Humboldt River include limestone, quartzites, and shales, which are intruded by a great variety of igneous rocks and overlain by basalts and rhyolites. No fossils have been found in this portion of the range, but on lithologic grounds the sedimentary rocks have been referred to the Carboniferous. The intruded rocks include granodiorites, quartz monzonites, quartz diorites, dacites, and andesites. It has been suggested that the granular rock of Agate Pass[b] is probably Archean, but the granitic rock in Mill Canyon and at Tenabo Peak is certainly later than the sedimentary rocks called Carboniferous, for the latter are metamorphosed at the contact. Neither the granular nor the sedimentary rocks have the schistose structure commonly found in pre-Cambrian formations.

The sequence or relative age of the eruptive rocks in this part of the range has not been determined, but probably all are later than the Carboniferous. The granular rocks are older than the rhyolites and basalts. On the east flank the range is bordered by a wide belt of the Humboldt formation which trends southward from Palisade for about 30 miles. Above the Humboldt and extending high on the east slopes of the range, at some places almost to the summit, there is a wide belt of basalt. The rocks along the crest of the western slope of the range from the Safford district to Mill Canyon include limestones, quartzites, granodiorites, diorite, rhyolites, and dacites. The prevailing dip of the bedded rocks is eastward.

CORTEZ AND MILL CANYON DISTRICT.

LOCATION AND HISTORY.

Tenabo Peak, near the southwest end of the Cortez Range, is about 30 miles south of Beowawe, the nearest station on the Southern Pacific Railroad. Cortez, on the southwest slope of this peak, was

---

a U. S. Geol. Expl. 40th Par., vol. 2, 1877, p. 570.   b Idem, p. 575.

one of the most productive mining camps in this part of Nevada. There are also a number of mines and prospects in Mill Canyon on the north slope of Tenabo Peak and on the slopes of Bullion Hill at the head of Mill Creek. The mines of Cortez and Mill Canyon were discovered in 1863, and rich ore yielding several hundred dollars to the ton was hauled to Austin for metallurgical treatment. Simeon Wenban, one of the early locators, obtained control of the Garrison, St. Louis, Arctic, Fitzgerald, and other important claims and organized the Tenabo Mill and Mines Company, which operated the mines for a number of years and still owns them. A small stamp mill with roasters and silver pans was built in Mill Canyon in 1864 and for three years was employed in treating the ore from various mines near by. This mill was bought by Wenban in 1867 and the ore was hauled to it from Cortez by wagon and pack trains.

The mines on the Cortez side of the mountain were exploited by Wenban with some success for many years. In 1886 a leaching plant with a daily capacity of about 50 tons was built at Cortez and this plant was in constant service for eight years. Water was piped from the mountains to the south, a distance of 7 miles, and the fuel supply was pine timber, a scattered growth of which is found on the mountain slopes. From the best sources of information that are available the production of Cortez and Mill Canyon since discovery is estimated at $10,000,000, the larger part of which was taken from the Garrison mine.

In the summer of 1908 cyanide tanks were built to treat the tailings from the Cortez mill, which are estimated to amount to about 120,000 tons. At this time the mines were not producing.

GEOLOGY.

Tenabo Peak is composed of limestone and quartzite which are intruded by granitic rocks and porphyries. The southwest slope of this mountain is one of the most striking features of this portion of Nevada. High on the slope a massive ledge of quartzite from 200 to 300 feet thick forms an abrupt wall, which for a distance of about 2 miles is practically vertical. This bed strikes a few degrees west of south and dips about 23° E. On the southeast slope of Tenabo Peak it descends steeply toward the valley flat and is covered by flows of Tertiary lavas. Underneath the quartzite is a series of gray limestones, presumably the Pogonip, which are conformable with the quartzite in dip and have a thickness of not less than 2,000 feet. At the top of the mountain above the massive ledge is a series of gray limestones, possibly the Lone Mountain, which are cherty near the base and pass upward into thinly bedded limestones with some siliceous bands. The relations are indicated by figure 18. The

massive wall of quartzite appears to be uninterrupted by faulting
except at the St. Louis mine, where a short block of ground is faulted
above the ledge about 100 feet. On the northwest slope of the

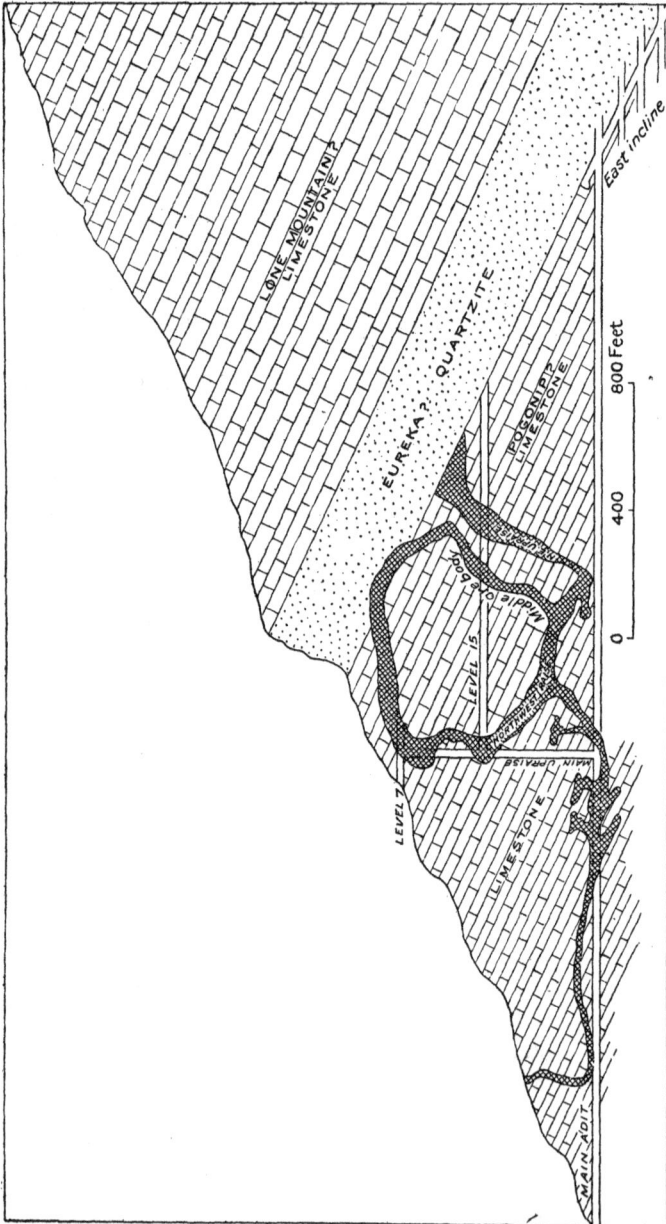

FIGURE 18.—Section drawn eastward through Tenabo Peak, showing underground workings at Garrison mine. The position of the ore chambers is taken from the maps of the Tenabo Mill and Mines Company, and some of the ore bodies are projected on the plane of the section.

mountain a great granite intrusive cuts across the sedimentary beds
and forms a broad precipitous spur which is cut by a number of
ravines that drain westward to Crescent Valley. This granitic mass

extends northeastward for several miles, forming the summit of the Cortez Range to the north of Tenabo Peak and occupying its western slope. In Mill Canyon, where the granitic mass is in contact with the sedimentary rocks, it sends out small apophyses into the limestone. In some places tremolite is formed in the country rock near the contact, but garnet zones seem not to have been developed. The granitic intrusive varies in composition from a rock composed of quartz feldspar and only a little biotite to one in which the dark silicates are present in considerable quantity. A thin section of a specimen taken from the western slope of Tenabo Peak and from the upper portion of the mass is composed of quartz, orthoclase, oligoclase, biotite, and muscovite, and is therefore quartz monzonite. In Mill Canyon a more basic phase of the intrusive rock contains considerable hornblende and less orthoclase and may properly be called granodiorite.

CORTEZ MINES.

Garrison mine.—The mines on the south and west slopes of Tenabo Peak belong to the Tenabo Mill and Mines Company. The Garrison mine, which is the most extensively developed, is located about a mile northeast of Cortez and in it centers a labyrinth of closely spaced intersecting workings from which a number of long crosscuts are driven to prospect the various claims near by. The level that is most extensively developed has an elevation of about 7,000 feet above sea level. A tunnel is driven eastward for 4,000 feet and from this several long crosscuts are run to the south. At 1,600 feet from the portal a raise inclined 85° N. connects the main tunnel with the seventh-level adit, and from this raise six levels are turned at unequal vertical distances. On the main adit, 1,500 feet east of the raise and just below the quartzite, an incline is driven eastward at an angle of 23° to a depth of 145 feet below the adit level, and three other short inclines are driven from crosscuts on the adit level. On level 5 the Garrison mine is connected by a long crosscut with the St. Louis mine and on level 6 with the Fitzgerald, but these connections had caved when the mines were visited.

Nearly all the workings of the Garrison mine are in limestone and below the quartzite cliff which forms the escarpment of the mountain. The usual strike is S. 11° W. and the dip is 24° to 27° E. Although there is some minor warping and puckering of the strata, their attitude is fairly constant over a considerable area.

A dike of decomposed igneous rock cutting through the limestone strikes S. 75° E. and dips northward at a high angle. On the adit level this is exposed here and there for 2,000 feet or more from a point west of the main upraise to the east incline. Three dikes, similarly decomposed and with approximately the same strike and dip, are crosscut in the Boss tunnel on the adit level and small exposures of the same rock may be noted at several places on the surface above the mine. These dikes, which are everywhere highly altered, are now composed mainly of quartz, sericite, and calcite; the original mineral constituents can not be made out. The dike, which is exposed here and there along the main adit, is not continuous, and from the lack of continuity it presumably did not everywhere fill the fissure into which it was intruded but was injected here and there from below. At some places along the dike the limestone near the contact is metamorphosed to a rock composed of quartz, calcite, tremolite, actinolite, orthoclase, pyrite, sericite, and biotite. In certain phases enough sulphides are present to constitute an ore, but the developments along the dike, though they include altogether several thousand feet on the six levels, have not been profitable. After the dike solidified it was sheeted and shattered, the fissures being in a broad way parallel to the intrusive but extending into the limestone wall rock also. These fissures were filled with banded sulphide ore, but most of those which are exposed on the lower levels are of too low grade to work.

The principal deposits of the Garrison mine are very irregular chambers of ore in limestone. The minerals are quartz, calcite, galena, stibnite, pyrite, zinc blende, stromeyerite, gray copper, and other minerals containing antimony and arsenic. The oxidized ore is composed of silver chloride, copper carbonates, and iron and manganese oxides. The galena is very rich in silver, especially where it is coated with dark films of a sooty black sulphide that is presumably argentite. The ore in mill runs ranges from 30 to 80 ounces of silver and $3 in gold to the ton. Where pyrite is abundant the mill runs range up to $15 in gold to the ton. The completely oxidized ores were very much richer. According to J. D. Hague,[a] 88 tons of ore taken from the St. Louis mine in 1868 yielded $600 to the ton.

The ore body which led to the discovery of the mine outcropped at a place near the top of an air shaft, which is raised to the surface, 400 feet in from the portal of the main adit. From the point of discovery the ore, which carried galena and silver chlorides, was followed almost vertically along a vein striking N. 60° E., to a point near the tunnel level, where the ore made out in irregular chambers in the limestone. A long, flat ribbon of ore was followed not far above the tunnel level to the Red Breast stope, a distance of about 700 feet. From the top of the Red Breast stope the ore dipped about 12° S. and was followed down a low-angle incline for 275 feet, to a point 60 feet below the tunnel level.

Eastward, along a fissure which strikes N. 20° W. and dips 55° NE., this ore joined the northwest ore body, which dips about 38° SE., as

a U. S. Geol. Expl. 40th Par., vol. 3, 1870, p. 406.

shown in figure 19. The northwest ore body is a ribbon of ore 350 feet long and from 25 to 100 feet wide, which cuts across the limestone. As shown by the maps of the company and indicated in figure 18, it extended upward almost vertically from level 5 to level 7, forming a large flat-lying mass, which made against the quartzite roof. Farther east the middle ore body, which was approximately in the same plane, joined the large mass near the base of the quartzite ledge. At the Capps breast, about 800 feet southeast of the main upraise, an ore body which pitches westward at a high angle is followed upward through very tortuous workings to level 5.

The fissure veins that occur in the zone of sheeting parallel to the dike are stoped here and there on the tunnel level and on two levels driven from the east incline, shown in figure 18. The large stopes

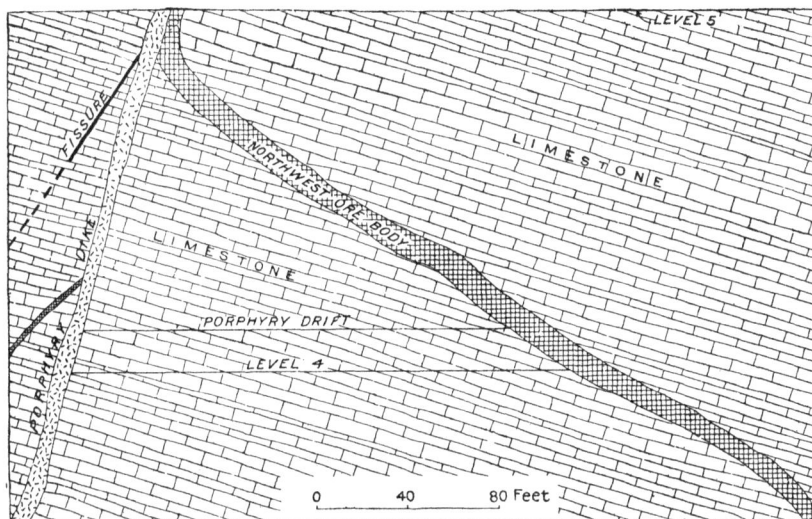

FIGURE 19.—Section through Garrison mine, Cortez. Line of section trends N. 15° W. along northwest ore body, about 10 feet north of main winze.

above the fifth level, which were inaccessible when the mine was visited, are said to be in the plane of the sheeted zone. They are above the intersections of this zone and the ore bodies, which in the lower part of the mine are called the northwest and northeast ore channels.

On the south and west slopes of Tenabo Peak there are several smaller mines which the Tenabo Company exploited many years ago. On the Arctic claim, above the wagon road between the Garrison mine and the mill, a short incline is driven down a bedding-plane deposit, making off from a narrow fissure vein which cuts across the bedding of the limestone. Figure 20 is a cross section of this deposit. On the mountain trail between Cortez and Mill Canyon there are a number of small deposits of highly siliceous silver ore in limestone and in quartzite.

Valley View mine.—The Valley View mine is 3 miles northwest of Cortez, on the border of the Crescent Valley. The country is an area of tilted limestone overlain by rhyolite and cut by porphyry dikes similar to those of the Cortez mine. Between Cortez and the Valley View mine are numerous small outcrops of ore showing copper carbonates, and in the vicinity of the Valley View are several small veins of lead ore. The Valley View vein follows in a general way a dike of altered porphyry. A shaft is sunk on this vein at an inclination of 58° to a depth of 103 feet, and at a depth of 50 feet a level is turned on the vein for 50 feet to the east; at the bottom of the shaft a second drift is run for 100 feet to the east. The vein is from 1 to 5 feet wide, strikes southeastward, and consists of white quartz, galena, pyrite, and silver chloride. The values are in lead, silver, and gold.

White Horse turquoise mine.—About half a mile south of the Valley View mine some small open cuts show a white decomposed rock which is presumably an altered rhyolite. This rock contains many small

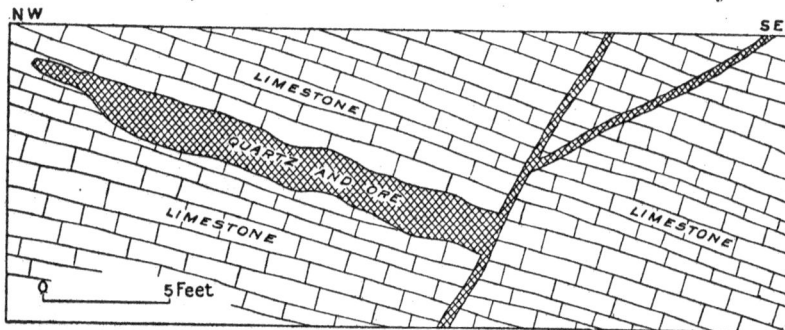

FIGURE 20.—Cross section of Arctic mine, Cortez district, near top of incline, looking north.

veinlets of turquoise from one-sixteenth to one-quarter of an inch wide. Some good gem material is said to have been obtained from this mine.

MILL CREEK MINES.

General features.—Mill Creek, the largest of several small streams which flow northwestward from the crest of the Cortez Range, drains the northern slope of Tenabo Peak through a steep-walled canyon. The mouth of this canyon is deeply incised in limestone, which about a mile above is intruded by granodiorite. To the northeast of the canyon are massive beds of quartzite and near the head are limestones that are presumably of the same age as the beds which form the summit of Tenabo Peak. Although the veins on this side of the mountain have not been so productive as the mines near Cortez, the country is highly mineralized and there is a great variety of deposits, some of which have produced considerable ore. The ore bodies are fissure veins of silver ore in granodiorite, silver-lead replacement deposits in limestone, and ferruginous deposits of gold ore replacing limestone.

The fissure deposits, which are the most important of the canyon, include the Aurora, Benjamin Harrison, Rhoda, Empire State, and other veins on Bullion Hill, a lofty spur which extends northward from Tenabo Peak. Most of the veins strike a few degrees west of north and dip eastward at moderately low angles. The ore consists of banded quartz and sulphides and carries from 30 to 200 ounces of silver and considerable lead. The walls along the veins are strongly leached by the hot waters which deposited the ores. Pyrite, sericite, calcite, and quartz are formed in abundance, replacing the ferro-magnesian minerals and feldspars. The minerals of the ore are quartz, calcite, galena, zinc blende, pyrite, argentite, stephanite, gray copper, and stibnite. Along the outcrops and in some places as deep as 100 feet below the surface the veins are altered to spongy ferruginous quartz carrying silver chloride, lead carbonate, and other minerals.

The silver-lead deposits in limestone occur in the main within 300 feet of the intrusive granodiorite and are of the irregular replacement type, forming in the limestone along zones of fissuring and at some places following the bedding of the limestone. The minerals are calcite, quartz, galena, pyrite, zinc blende, and chalcopyrite. The values are mainly silver, but some of the deposits carry considerable gold.

Bullion Hill mines.—The Aurora vein of the Bullion Hill group strikes a few degrees west of north and is opened here and there on the surface for a distance of about 2,000 feet. The Water tunnel on this vein follows it for about 600 feet along the strike. The lode dips eastward and in places is a sheeted zone from 6 inches to 5 feet wide, with slabs of country rock between the ore shoots. The Aurora tunnel, 375 feet long, is driven in decomposed granodiorite 58 feet above the level of the Water tunnel, and part of the workings are west of the Aurora vein. Here and there stopes are carried up from the Water tunnel to the level of the Aurora tunnel. The ore minerals are quartz, calcite, galena, zinc blende, pyrite, chalcopyrite, gray copper, and a sooty black powder on galena, which is probably silver glance. A little ruby silver was found in the bottom of the winze on the Water tunnel level. The vein is 2 feet wide and carries about 75 ounces of silver to the ton. It has produced $50,000 since 1883, when it was first located.

The Rhoda fissure parallels the Aurora vein and lies 30 feet to the east on the Water tunnel level. It is a great zone of crushed granite with rounded friction fragments of granite and quartz. The Rhoda shaft, which is now inaccessible, was put down 250 feet on an incline 80° W. A shoot of rich ore found in this shaft is said to have been cut off by a fault, which is possibly the same as one exposed in the Water tunnel 30 feet east of the Aurora vein. The Bullion tunnel, 260 feet below the Water tunnel level, has been driven for about 1,000 feet westward to tap the Rhoda and Aurora veins. Several small veins parallel to the Aurora have been found, but nothing of great value has yet been discovered in this tunnel.

The Benjamin Harrison vein, which lies to the east of the Aurora vein, on an adjoining claim, has been developed in tunnels and pits for a distance of about 2,000 feet. This vein is in quartz monzonite, strikes N. 25° W., and dips toward the east. The ore and the alteration of the wall rock are similar to those of the Aurora, but the vein is narrower. It belongs to the Bullion Hill Company and is said to have produced $8,000 in silver.

Empire State mine.—The Empire State mine, one-fourth mile above the Mill Canyon mill, was located in 1872 and has produced about $2,500. The deposit is a fissure vein in granodiorite that strikes S. 55° W. and dips 35° SE. The lode is exposed in three tunnels, the highest one about 100 feet above the lowest. In the lower tunnel the vein, which is about 2 feet wide, is stoped here and there at several places and is offset by small normal faults which dip southwestward. About 50 feet from the face a fault nearly parallel to the vein offsets it a few feet toward the west. The ore is somewhat similar to the Bullion Hill ore, but contains less galena and has higher gold values. The minerals of the sulphide ore are pyrite, zinc blende, galena, stibnite, and gray copper.

Hidden Treasure mine.—The Hidden Treasure mine is one-half mile S. 30° E. of the Rhoda shaft of Bullion Hill. From a short tunnel a 60-foot incline is driven S. 23° W. at an angle of 25°, on a lode of highly oxidized iron-stained gold ore. The hanging wall is limestone and the foot wall a white decomposed igneous rock which is probably leached quartz monzonite. At the top of the winze the ore is several feet wide and carries $35 in gold to the ton. It thins out at the bottom of the incline, but the face of a drift along the contact for 75 feet west from the bottom is in silicified iron-stained limestone, which is said to carry $20 in gold to the ton. At 150 feet west of the portal of the tunnel a second incline 70 feet deep is driven in limestone and intersects the same contact at the bottom.

Lewis Canyon claims.—Near the contact the limestone carries bunches of oxidized gold ore. In Lewis Canyon, on the west slope of Bullion Hill, granodiorite cuts through limestone and sends off dikes into it. A number of fissure veins strike northward and dip from 50° to 80° E. They are in granodiorite, in limestone, and at contacts between the two rocks. The veins are up to 4 feet wide, are composed of quartz, calcite, galena, and pyrite, and carry values in silver. At the Isaacs mine a silver-bearing vein strikes north, dips 80° E., and is opened in several short tunnels. In the only accessible tunnel the vein is 3 feet wide and is composed of ribbons of quartz and sulphides forming bands parallel to the walls.

At the Cynthia mine, on the northwest slope of Bullion Hill, a lode which in places is 4 feet wide strikes south and dips 68° E. It is opened by short tunnels, shafts, and underhand stopes for several hundred yards along the strike. The ore is composed of quartz, calcite, galena, and pyrite and is said to carry 60 ounces of silver to the ton. The vein was one of the first discoveries in the camp and has produced a considerable quantity of ore, which in the sixties was sent to the Canyon mill.

Falconer and Berlin mines.—About one-fourth mile below the Canyon mill the contact of limestone and granite strikes N. 53° E. and dips northwest at a high angle. The limestone, which lies northwest of the granodiorite, is locally rather closely folded and in general dips away from it at a moderately steep angle. The granodiorite sends out small dikes, which cut across the bedding and form thin sheets in the bedding planes. Along the contact, extending for a

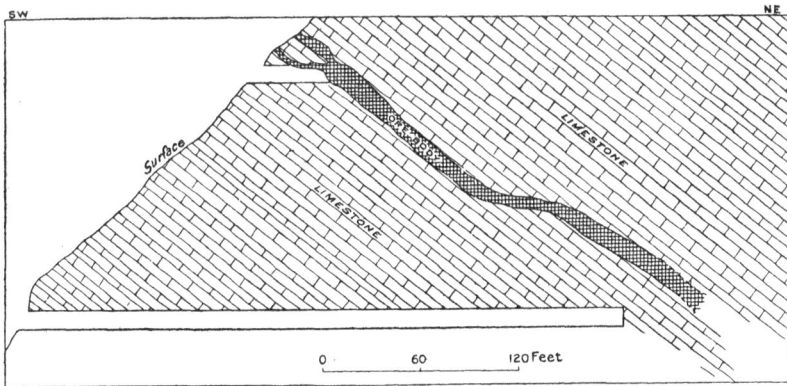

FIGURE 21.—Cross section of ore body of Falconer mine, Mill Canyon.

distance of several thousand feet, a number of small veins and irregular shoots of silver-lead ore are developed in the limestone within a few yards of the granite, and there is an almost continuous chain of small deposits up the canyon sides from the bottom of the gulch. The ore shoots are not large, but some of them are said to have produced ore of very high grade.

The Falconer mine is on the northeast side of the canyon about 300 feet above Mill Creek. The country rock is a dark-gray limestone which strikes S. 60° E. and dips 30° NW. The ore body is a replacement deposit that cuts the bedding at a small angle and here and there small masses follow bedding planes and the small fissures which cross them to join the main ore body. Figure 21 is a cross section of this mine. From a 40-foot tunnel near the apex of the ore a 120-foot incline is sunk on the ore and small stopes are carried up from the incline at several places in the deposit. At the bottom of the incline the ore shoot flattens and dips about 20° E. It is followed in a drift for 40 feet along the strike, at the end of which a winze is sunk 40 feet on a dip of 30° N. A crosscut tunnel 125 feet below the upper tunnel is driven northeastward for 375 feet. At 330 feet from the portal a raise is turned to the ore body to connect with the incline from the upper tunnel. The minerals of the ore are quartz, calcite, galena, pyrite, chalcopyrite, and zinc blende. Much of the ore is a black sooty powder, probably pyrite in the main. The ore is said to carry $35 to the ton in silver, lead, and gold.

The Berlin mine, on the southwest side of the canyon, has supplied about $40,000 in gold, silver, and lead. Most of the ore bodies are small and some seem to be related to fissures which strike northeastward and dip flatly to the southwest.

On the New York claim, which joins the Berlin on the northwest, a shaft is down 28 feet on a vein striking northwest and dipping 80° SW. At the face of a short drift this vein shows a foot of lead and iron sulphides which run high in gold. A car of ore has been shipped from this level. A 200-foot tunnel about 90 feet below the collar of the shaft has been run to intersect the ore below, but has not encountered it.

Caledonia mine.—The limestones on the northeast slope of Mill Canyon near its mouth strike N. 65° W. and dip 44° NE. A shoot of ore trending approximately with the bedding has been followed for 50 feet down an incline. A carload of ore taken from this incline and from a short drift from the bottom is reported to have yielded 120 ounces of silver and $20 in gold to the ton and 13 per cent of lead. Approximately 300 yards farther north, at about the same horizon in the limestone, two drifts are run on a thin bedding-plane deposit which is said to carry $25 in gold to the ton.

SAFFORD DISTRICT.

GENERAL FEATURES.

The Safford district includes a number of small mines and prospects situated in the vicinity of Barth, a small camp on Humboldt River about 6 miles west of Palisade. The district takes its name from James Safford, who discovered the Onondaga mine in 1881. The mines of the district have produced about $200,000 in silver, nearly all of which was taken from the Onondaga and Zenoli mines. The West Mining Company is exploiting an iron deposit at Barth and ships from 100 to 300 tons of iron ore daily to the Utah plants of the American Smelting and Refining Company, where the ore is used for flux.

In the report of the Fortieth Parallel Survey the rocks of the Safford district are represented as trachytes, which are said to overlie andesites. The rocks collected by the writer proved to be varied in composition and included fine-grained diorites that carry some orthoclase and quartz, glassy vesicular andesites, and dacite porphyries. These species are probably the eruptions or the intrusions from a center of Tertiary volcanism which has been deeply eroded by Humboldt River, but which in the higher country on both sides of the river is nearly everywhere surrounded by rhyolite. The deposits are fissure veins of silver ore, which carries among other minerals some antimony and arsenic compounds—a type of deposit which is common in the Tertiary eruptive rocks. The wall rock is somewhat altered near the veins, the feldspar having been partly changed to sericite and calcite, and the ferromagnesian minerals are nearly everywhere altered to chlorite. These changes, which are presumably the result of the action of the vein-forming solutions on the wall rock, are not so intense as the hydrothermal action at Tuscarora and at some other places in the later Tertiary lavas.

MINE DESCRIPTIONS.

Zenoli mine.—The Zenoli mine is about 1 mile southeast of Barth, .n a small gulch which joins the Humboldt Valley. The mine was discovered by Italian prospectors in the eighties and was held by them for a number of years, yielding a small amount of shipping ore. In 1907 the Zenoli Silver Copper Company was organized and since that time has been producing ore steadily and employing a force of about ten miners. The values are in silver, with important amounts of copper and lead, and the ore shipped carries from $60 to $70 to the ton. An inclined shaft sunk on the principal deposit to a depth of about 100 feet below the surface is intersected by an adit 220 feet from its portal, and levels are turned from this shaft at vertical intervals about 10 feet apart, making altogether about 2,500 feet of underground workings. The country rock is an andesite, which carries here and there a crystal of resorbed quartz and approaches dacite in composition. The deposit is a well-defined fissure vein from 1 to 5 feet wide. The sulphide ore is composed of quartz, calcite, barite, stibnite, gray copper, galena, pyrite, and chalcopyrite. In the upper levels there is considerable iron oxide, copper carbonates, and horn silver. The ore, which is sorted in the stopes, is shipped to Utah smelters. The principal lode strikes about north and near the surface dips 23° E.; on the 50-foot level the dip steepens to 45°. Nearly everywhere on the foot wall of the vein there is a smooth, well-defined slickensided surface carrying more or less gouge. On the tunnel level this plane of movement departs from the foot wall, cutting across the vein and displacing it to an undetermined extent. A second vein, known as the Lead Stope vein, has also a north-south strike, but dips eastward at a higher angle, joining the main lode near the surface. Some high-grade ore was found above the junction of the two veins.

Onondaga mine.—The Onondaga mine is located on the south side of the gulch, about 500 yards southeast of the Zenoli mine. Two lodes outcrop boldly on top of the hill. The principal one strikes northwestward and dips from 60° to 80° SW. It is developed for a distance of 700 feet by several short tunnels and underhand stopes driven downward from the grass roots, the workings having a vertical range of 250 feet. A tunnel is driven northwestward for 600 feet and intersects the bottom of a shaft 200 feet in depth about 565 feet from the portal. The country rock is andesite, which is altered near the vein, and the joint planes are covered with a dark coating that is probably a mixture of iron and manganese oxides. The ore consists of quartz, iron oxide, calcite, and barite, with copper carbonates and oxides, and is said to carry from 30 to 100 ounces of silver to the ton.

In the gulch below the mine on the Malachite claim a crosscut tunnel is driven northward for 350 feet toward the Onondaga vein.

Ruby claim.—The Ruby claim, which is a few rods east of the Zenoli mine, adjoins the Malachite claim on the northwest. Four tunnels, from 50 to 250 feet long, have been driven to explore three veins which strike northwestward. The principal lode is an iron-stained zone of crushed porphyry, which here and there carries banded ribbons of ore composed of quartz, chalcopyrite, pyrite, gray copper, and iron oxides, with a little ruby silver. A carload of selected ore, running about 100 ounces of silver to the ton, has been shipped to smelters.

Humboldt mine.—The Humboldt mine is located on the north side of Humboldt River, about 3½ miles below Palisade. Andesite, which is locally vesicular, outcrops at the surface, and copper-bearing sulphides have been deposited in a zone of brecciation. Some ore taken from a shallow pit shows bornite, chalcopyrite, and chalcocite, partly altered to copper carbonates and iron oxides. A crosscut tunnel 200 feet long has been driven at the level of the river, but has not yet encountered the lode.

Bonanza mine.—The Bonanza mine is about three-fourths of a mile S. 70° E. of the Humboldt mine. The lode is a zone of crushed and altered eruptive rock, cemented by iron oxides. In the shattered zone, which can be traced for about 200 feet, thin sheets of galena, rich in silver, may be found here and there at the surface, filling the crevices in the shattered rock. Two shallow pits have been sunk on the deposit.

Pittsburg and Palisade tunnel.—On the Pittsburg and Palisade group of claims, which join the Humboldt on the west, a crosscut tunnel has been driven for 600 feet and is intended to intersect a small vein that outcrops on the hill above.

West iron mine.—The West iron mine is located at Barth, 6 miles below Palisade. An open cut 200 feet in diameter and 80 feet deep is dug in the red hematite ore, which is shipped to Salt Lake and used for flux. The rock is quarried from the pit, hoisted up an incline, and loaded into railroad cars.

The cut is approximately the size of the known area of iron ore. The rock to the north and east is covered by gravels, which at the pit are about 20 feet deep, and on the south and west the iron ore is in contact with a fine-grained diorite. About 100 feet south of the pit the fine-grained diorite is in contact with an altered rock, probably a metamorphosed limestone. The iron ore, which is in the main red hematite, carries a very small amount of magnetite and specularite and is cut by many small veinlets of apatite. The deposit is presumably of contact-metamorphic origin and not the capping of a sulphide ore. The sulphides alter mainly to hydrated iron oxide, and not to hematite and magnetite, like the ore of the West mine. Outcrops of iron deposits are reported to have been found also in the Cortez Range about 15 miles south of Palisade.

SHOSHONE RANGE.

GENERAL FEATURES.

The Shoshone Range extends southwestward from Humboldt River near Beowawe for 50 miles or more. The present reconnaissance covered only the northern part of the range, which includes a group of lofty mountains, among them Shoshone Peak, 9,760 feet above the sea, the most elevated portion of the mountain mass. The central portion of the north end of the range is a group of steep-walled mountains with deeply incised amphitheaters, which are clearly the result of mountain glaciation. Northeast of the group of elevated peaks is the Whirlwind Mesa, which extends about 15 miles to Humboldt River. This mesa, which is over 7,000 feet above sea level, slopes gently northward and is divided by the Whirlwind Valley, which joins the Humboldt Valley near Beowawe.

As the central mountain mass is higher than the neighboring mountains, it is better watered and it supports a scattered growth of timber useful for local needs. The mining camps of the north end of the range are Tenabo, Lander, Mud Springs, Grey Eagle, Hilltop, Maysville, Pittsburg, Dean, and Lewis; all of these are on the east and north slopes of the mountains and are included in a rectangular area 6 by 16 miles. In the eighties there was considerable mining in some of these camps, especially at Lewis, Dean, and Lander. The Battle Mountain and Lewis Railroad, a narrow-gage line, was built from a point on the Nevada Central Railroad through Lewis to Quartz Mountain, at the north end of the range.[a] This road was never profitable and it was dismantled many years ago. At present the transportation to the various camps is by wagon road from Beowawe and from Battle Mountain.

---

a Bancroft, H. H., Works, vol. 25, Nevada, p. 238.

Development work was being carried on at several of these camps in 1908 and some of the mines were shipping a little ore. The Dean mill, the only one in working condition, was idle at that time.

GEOLOGIC FEATURES.

The most abundant rocks of the northern part of the Shoshone Range are quartzites, siliceous shales, and limestones. Interbedded with these are conglomerate beds which are exposed, among other places, on the north slope of Shoshone Peak. On lithologic grounds [a] the sedimentary series has been termed Carboniferous, but a considerable thickness of other rocks may be represented. At most places the sedimentary rocks dip toward the east or southeast from 20° to 50°, but locally they dip at high angles in other directions. The structure, as shown in the atlas of the Fortieth Parallel Survey, is that of an eastward-dipping monocline crossed by normal faults which dip westward. The sedimentary rocks are intruded by large masses of granodiorite and granodiorite porphyry. Such intrusive rocks are exposed at Tenabo, Mud Springs, Grey Eagle, Hilltop, and Dean, and the largest one of them forms the central part of Shoshone Mountain. An analysis of this rock, made by R. W. Woodward, is given on page 25, where the granodiorites are described.

Some actinolite and doubtless other contact-metamorphic minerals are developed in the quartzite, but no garnet zones were found in the few places where the contacts were observed. The siliceous sedimentary rocks are the most abundant and these, as is well known, are not so favorable to contact metamorphism as the calcareous rocks. Many of the ore deposits are in or near the granodiorite.

Dense andesites outcrop at Tenabo and at Lander, where they are surrounded by sedimentary rocks. The andesites have a glassy groundmass and in some places they are vesicular. It is assumed that they represent a magma which consolidated as an intrusive near the surface or as a flow. A broad belt of rhyolite borders the range low down on the east slope. A small intrusive mass of dense quartz porphyry was noted on the north slope of the gulch below the Pittsburg mine.

TENABO.

GENERAL STATEMENT.

Tenabo is situated on the east slope of the Shoshone Range, near the edge of Crescent Valley, about 21 miles southwest of Beowawe. It is in the Bullion mining district, which was organized many years ago to cover locations at Lander, but when a number of ore deposits were discovered in 1907 at some distance from that camp the new

---

a U. S. Geol. Expl. 40th Par., vol. 2, 1877, p. 619.

town Tenabo was established 2 or 3 miles away, more conveniently situated for them. The town is well laid out and has some substantial wooden buildings, but is without a regular water supply. Early in 1907, when there was a rush to the mines, about a thousand persons were living there, but within a year all except a few score had left. A stage is run thrice a week between Tenabo and Beowawe, and during times of prosperity automobiles and a steam traction line were operated. The name Tenabo is not well chosen, as it is likely to be confused with that of Tenabo Peak, on the opposite side of the Crescent Valley, where the Tenabo Mill and Mines Company carried on extensive operations for many years.

GEOLOGY.

The relief at Tenabo is slight and consequently the natural exposures are not so good as at camps which are situated in more commanding positions, but owing to the large number of location pits recently dug a fairly satisfactory knowledge of the geologic features may be obtained. The rock formation which occupies the greater part of the area is composed of quartzites and fine-grained siliceous shales that have a general eastward dip of 15° to 40°. There are no pure limestones in the area near Tenabo, but some of the sandy layers of the quartzitic formation contain calcium carbonate which cements the grains of quartz. Certain fissile layers on the ridge above the Gem mine are dark and very fine grained and in the field resemble carbonaceous shale, but under the microscope they are seen to be composed of very fine particles of quartz with a small amount of sericite between the grains. The series is termed Carboniferous in the reports of the Fortieth Parallel Survey. The sedimentary rocks are intruded by porphyritic granodiorite, which is best exposed in the Phoenix mine. This rock in hand specimen has a dense gray groundmass, in which are embedded phenocrysts of feldspar, quartz, biotite, and hornblende. Under the microscope the groundmass is seen to be composed of quartz and orthoclase crystals which are distinctly smaller than the phenocrysts but larger than the constituents of the groundmass of the ordinary porphyries. The feldspar phenocrysts are in the main oligoclase and andesine, and many of them are zonally built. At the Two Widows claim a quartz diorite porphyry, which is probably a phase of the granodiorite magma, intrudes siliceous shales. Near the contact small veinlets of actinolite are developed, presumably as a result of contact metamorphism, but in the sections studied no garnet or other contact minerals have been found.

A dark, dense rock with an altered glassy groundmass containing phenocrysts of orthoclase and acidic plagioclase is exposed at the Gem mine and on the hills above the shaft. This rock will be called andesite, but its determination is unsatisfactory owing to the decomposed condition of the material at hand; the slides studied contain more orthoclase than the other andesites described in this report, and possibly the rock should be classed as latite. It is vesicular in places and is everywhere finely crystalline and obscurely porphyritic, and it is therefore probably a flow or an intrusive formed near the surface. The andesite is believed to be younger than the other porphyries and appears to have formed at less depth. Near the lodes the country rock is hydrothermally altered, but the altered rock is restricted to a relatively narrow zone. Sericite, calcite, and pyrite have formed in the granodiorite, and chlorite, sericite, and pyrite are developed in the andesite.

ORE DEPOSITS. .

The ore deposits are sheeted zones in quartzose sedimentary rocks and in andesite and fissure fillings in granite. The lodes do not fall into well-defined groups or parallel systems, but dip to all points of the compass. On the surface the ore is iron-stained quartz, carrying some chloride, and is usually kaolinic. Some of the surface ore gives very high pannings in gold. Sulphides appear in depth from 50 to 100 feet below the surface, and include arsenopyrite, pyrite, chalcopyrite, bornite, galena, zinc blende, and chalcocite. The gangue is quartz and calcite. Thin films of molybdenite appear on the joint faces of the country rock in the Violet shaft. Since the deposition of the ore there has been considerable movement, and practically all the lodes are sheared by slickensided planes parallel to the vein or are faulted by cross faults. These are small as far as developed and do not cause loss of the vein in any of the mines visited.

MINE DESCRIPTIONS.

Little Gem mine.—The Little Gem mine, located about 1½ miles west of Tenabo, was discovered in 1907 and is the most extensively developed property in the area. A shaft is driven southwestward on the lode at inclinations of 20° to 30°.. This shaft is 400 feet long and on the four levels turned from it there are altogether about 900 feet of drifts. In 1907 eighteen carloads of ore was shipped and yielded about $30 a ton.

The country rock is a dark, fine-grained andesite, showing very few phenocrysts even where most porphyritic. On the hill just above the mine it is vesicular. The andesite is strongly sheeted parallel to the lode and is altered near the lode, but at most places it is not strongly leached; chlorite and pyrite, with some sericite and calcite, are developed by secondary processes, chlorite being formed in the greatest abundance. The ore outcrops near the collar of the shaft and at the outcrop is composed of iron oxide and quartz carrying copper carbonates, silver chloride, and free gold. The sulphide ore, which appears about 75 feet below the surface, is composed of quartz, arsenopyrite, pyrite, chalcopyrite, galena, and zinc blende, with a little bornite and a sooty black film which covers other sulphides and is probably chalcocite. On level 1 the lode strikes north of west, but on the second level, the most extensive in the mine, it bends and strikes south of west. It dips about 30° S., a little more steeply than the incline, which is driven on the lode but to the right of the line of steepest dip. The lode is from 1 to 6 feet wide and, in the main, is a zone composed of several closely spaced parallel sheets of quartz and sulphides, between which the country rock is highly fractured and seamed with veinlets crossing the general strike of the vein. Thin drusy cavities parallel to the walls are lined with sulphide coatings, and these are in turn covered with quartz crystals pointing to the center. Since it was formed the vein has been much crushed and shattered by movement, which in the main was parallel to the walls. It is also crossed by several small normal faults, the offsets of which are not great enough to throw the vein out of continuous view in regular workings. The mine is reported to have in sight 7,000 tons of ore, with an average value of $2 in gold and 10 ounces of silver to the ton and 3 per cent of copper.

Phoenix mine.—At the Phoenix mine, 1¼ miles southwest of Tenabo, a two-compartment vertical shaft is 250 feet deep, and an adit driven northward for 280 feet intersects the shaft about 100 feet below the collar. This adit is continued northward for 220 feet beyond the shaft, and short drifts are run on two veins at this level. A second level is turned from the shaft 60 feet below the adit. The country rock is porphyritic granodiorite, cut by a mass of intruding quartz porphyry, which measured on the adit level is about 100 feet thick. The south contact of the quartz porphyry, where it is encountered in the main adit, 70 feet from the portal, dips about 20° S. Three veins are exposed, two of them in granodiorite and one at the north contact of the quartz porphyry and granodiorite. They are from 4 to 18 inches wide and all strike westward. The deposits are banded fissure fillings, and the country rock near the lode is strongly sericitized. The sulphide ore, 100 feet in depth, is composed of quartz, pyrite, arsenopyrite, chalcopyrite, galena, and blende. In the granite sericite, calcite, chlorite, and pyrite have been deposited by ore solutions. There has been much movement since the veins were formed, and in places the lodes consist of 2 feet of white sericitized decomposed granite, containing here and there broken masses of quartz and ore up to a foot thick.

Gold Quartz mine.—The Gold Quartz mine is 700 yards south of the Little Gem. A vertical two-compartment shaft is down 308 feet, with short levels turned at intervals of 35, 100, 150, and 275 feet below the collar.   The country rock is quartzite, cut by an intruding quartz-bearing porphyry and by basalt.   In a shallow opening near the main shaft some rich gold ore was encountered,   From this and from the main shaft a shipment of ore, with an average value of $80 a ton in gold, has been made.   The ore is highly oxidized; about 75 feet below the surface the sulphides appear, arsenopyrite and pyrite predominating.   In the main shaft, between the 35-foot and 100-foot levels, there is a body of shattered, altered quartzite, which is cut by stringers of iron oxide carrying free gold.   This ore body is 4 feet wide, strikes N. 20° W., and dips 35° NE.   Between the 35-foot and 100-foot levels an incline follows it for 30 feet; at the bottom of the incline the lode breaks into several small stringers, and this deposit has not been encountered below the 100-foot level.

Violet claim.—The Violet shaft, about 1 mile southwest of Tenabo, is down 208 feet in siliceous shale and quartzite.   A crosscut is being run below an outcrop which has been opened in a surface pit. In the jointing of the country rock there are some small seams of molybdenite.

Two Widows claim.—On the Two Widows claim, half a mile west of Tenabo, an incline is sunk 110 feet at an angle of 70°, and from the bottom short crosscuts are run east and west.   The country rock is fine-grained quartzite cut by intruding quartz diorite porphyry. Veinlets of actinolite cut the quartzite near the intruding porphyry. The joint planes of the quartzite are filled with copper carbonates and iron oxides, which are said to carry 40 ounces of silver and several dollars in gold to the ton.

LANDER.

GENERAL FEATURES.

Lander, 2 miles northwest of Tenabo, is the oldest camp in the Bullion district and was the milling center for the district in the seventies and eighties, when the Lovie mines were being worked. The rocks of the area are siliceous and carbonaceous shales interbedded with quartzite and limestone and have a general southeastward dip.   The sedimentary rocks are capped with andesite and cut by intrusive quartz porphyry.   The principal mines are the Bonnie Jean, the Silver Prize, and the Silverside.

MINE DESCRIPTIONS.

Bonnie Jean mine.—The Bonnie Jean, known also as the Lovie mine, is 1½ miles northwest of Lander and is opened by four tunnels which have a vertical range of 250 feet.   In the gulch below these workings a shaft was being sunk in 1907 to explore the vein in depth. This mine was worked in a small way for many years, and the ore was treated in the 5-stamp pan-amalgamation mill at Lander.   It is currently reported to have produced some $300,000 in silver, the chloride ore mined near the surface of the deposits running several hundred ounces to the ton. The ore body is a fissure vein in siliceous shales and quartzite. Near the apex and northwest of the deposit is an outcrop of andesite similar to that of the Little Gem mine at Tenabo. The lode, which lies approximately with the bedding, strikes N. 65° E. and dips 20° to 50° SE. The outcrop of the deposit is exposed at the surface for about 400 feet along the strike and has been followed down the dip for about 300 feet. The oxidized ore is composed of quartz, iron oxides, silver chloride, and lead and copper carbonates. The values are principally in silver. There is a considerable tonnage of low-grade silver ore blocked out which the present owners consider concentrating ore.

Silver Prize vein.—On the Silver Prize claim, half a mile north of the Bonnie Jean, the country rock is decomposed andesite, and to the northwest, near the crest of the hill, are outcrops of massive quartzite. The lower vein, which is opened in three short tunnels, strikes northwest and is approximately vertical. The lode locally carries 2 feet of highly oxidized ore, and galena and zinc blende appear in depth. Higher on the hill and about 75 feet to the northeast of this lode a second lode strikes N. 50° W. and dips 58° SW. A 40-foot incline is driven on the lode, which carries 20 inches of rich silver-bearing galena and lead carbonate. The two veins probably join on the slope of the hill.

Silver Side mine.—At the Silver Side mine, in Lander, a tunnel is run 625 feet southward on a vein which dips 20° to 38° E., and from this stopes are carried here and there above and below. The vein, which is about 3 feet wide, is at most places approximately parallel to the bedding. The deposit with respect to its structural features resembles that of the Bonnie Jean, but the ore is not so rich.

MUD SPRINGS.

GENERAL FEATURES.

Mud Springs is on the east slope of the Shoshone Range, about 4 miles north of Lander. In the summer of 1907 a number of claims were located, and recently a few tons of ore has been shipped. The country rock consists of quartzite, with fine-grained, shaly beds, and is cut by granodiorite and andesite. The deposits are fissure veins in the sedimentary rock and in granodiorite. The minerals of the ore are quartz, limonite, lead carbonate, silver chloride, and gold.

MINE DESCRIPTIONS.

Triumph mine.—At the Triumph mine a narrow lode carrying oxidized gold and silver ore cuts across siliceous shales, strikes N. 60° W., and dips 45° SW. It has been followed for 40 feet down an incline from which a 50-foot drift has been run, and from this a stope has been raised east of the incline. On the surface 300 yards farther west a vein which is presumably the same cuts through a mass of undetermined igneous rock, probably an altered granodiorite. From this mine in the summer of 1908 a shipment of 18 tons of ore was made. This ore carried silver, lead, and gold, and is said to have yielded $60 a ton. The lode is from 1 to 3 feet wide, and the ore minerals include iron oxide, lead carbonate, and silver chloride.

Big Bug claim.—At the Big Bug claim two tunnels, each about 100 feet long, are driven to intersect a lode cutting across quartzite which dips steeply southeastward. The lode is about 6 inches wide, strikes N. 55° W., and dips 80° S. The decomposed iron-stained quartzose ore carries low values in gold and silver.

Bridal Wreath claim.—On the Bridal Wreath claim, which joins the Big Bug claim on the northeast, a 30-foot shaft is sunk on a crushed vein in quartzite which strikes S. 65° W. and dips 70° N.

Uncle Sam claim.—The Uncle Sam shaft, half a mile below the Big Bug, is sunk 50 feet on a silver lode which dips 76° S. A small vein of quartz and sulphides is located near a contact of granite and quartzite.

GREY EAGLE MINE.

The Grey Eagle mine, a mile or two west of Mud Springs and about 22 miles in an air line southeast of Battle Mountain, is located near the summit of a high granodiorite ridge, which is separated by a deep valley from the main axis of the Shoshone Range. The mine was worked in the seventies and eighties, lay idle some twenty years, and was worked again in 1905. It produced about $25,000 in 1906 and 1907. The main shaft is down 250 feet, and levels are turned at 60, 115, and 215 feet in depth. When the mine was visited in 1908 only the 60-foot level was accessible. The deposit is a fissure vein of banded quartz and sulphide, strikes N. 70° E., and dips 70° N. The country rock is a coarse granodiorite composed of oligoclase, andesine, quartz, hornblende, and biotite. Near the vein the granodiorite is strongly altered by hydrothermal metamorphism. The feldspars, biotite, and hornblende are replaced by sericite and by numerous small crystals of pyrite. The minerals of the ore include quartz, zinc blende, galena, pyrite, and gray copper, which near the surface are altered to oxides, carbonate, and chloride. The mine is reported to have produced since its discovery several hundred thousand dollars in silver, gold, and lead. It is commonly regarded as a property which possesses considerable promise.

HILLTOP.

Hilltop, or Marble Canyon, is a small camp which has recently been established on the east fork of Rock Creek, about 18 miles southeast of Battle Mountain. The country rock is quartzite, which includes fine siliceous shales with here and there some fine conglomerates containing small fragments of angular jasper. The sedimentary rocks are cut by dikes of green chloritic porphyry which under the microscope shows phenocrysts of quartz, orthoclase, and some oligoclase with ferromagnesian minerals altered to chlorite and sericite. This rock is presumably an altered granodiorite porphyry.

The principal mines are the Independence and Hilltop, which were located in 1906. Two 50-foot shafts and two short tunnels are driven on a zone of fractured quartzite, which trends northward and is exposed at intervals for about 2,000 feet. Small intrusive masses of leached porphyry cut this zone of quartzite at three or four places, the intrusive rock including fragments of the quartzite. At some places the shattered quartzite is seamed with stringers of quartz and iron oxide carrying a large amount of free gold. Here and there are small bodies of pyrite and galena in the fracture planes of quartzite and minute stringers of gray metallic mineral which is said to contain bismuth. Assays show the presence of silver, by weight equal to the gold. The quartzite near the ore is impregnated with pyrite and is locally decomposed to a soft white claylike substance which is said to carry high values in gold.

When the camp was visited the amount of high-grade ore in sight was small and developments were not sufficient to show the extent of mineralization in the fractured zone.

MAYSVILLE MINE.

The Maysville mine is on the east slope of Shoshone Peak about 1 mile west of Hilltop. The country rock is quartzite, which near the lode dips east of south about 35°. The principal vein cuts across the quartzite, striking N. 80° W. and dipping 80° S., and is opened by three short tunnels and a shaft sunk on the deposit. In places it is stoped to the grass roots, the surface ore consisting of yellowish-brown siliceous rock which carries silver chloride. The ore about 50 feet below the surface is composed of quartz, pyrite, galena, chalcopyrite, and gray copper. The vein, where accessible, is about 2 feet wide and shows banded quartz and pyrite parallel to the walls. The quartz and sulphides are highly crushed and at places mixed with much pulverized quartzite. About 1,500 feet northwest of the shaft a crosscut tunnel is driven southward for 215 feet to a vein, which is presumably the same lode. From this tunnel drifts are run 150 feet east and 75 feet west. The lode here strikes N. 60° W. and dips southward at a high angle. On the east drift some stoping has been done, but at most places the vein is small. A four-pan silver mill in the canyon below the mine was used some thirty years ago to treat the ore.

LEWIS AND DEAN.

LOCATION AND HISTORY.

Lewis is situated about 12 miles southeast of Battle Mountain, on the north edge of the elevated portion of the Shoshone Range, at the mouth of Lewis Canyon. Dean is 4 miles above Lewis, in the bottom of Lewis Canyon, which at this place is one of the most rugged and picturesque mountain features in northeastern Nevada. Silver deposits were discovered in Lewis Canyon in the seventies, and elaborate preparations were made by a company well provided with capital to develop the Eagle, Starr Grove, and other mines. A railroad which has long been dismantled was built from Lewis Junction to Lewis, and afterward was extended up Lewis Canyon to the Starr Grove mine. At Lewis a small silver amalgamation mill was built on the east side of the canyon, and considerable silver ore is said to have been put through this plant. Subsequently a 40-stamp mill equipped with roasters, pans, and settlers was built on the west side of the canyon, but was not long in operation. Extensive development work was done in several mines and some ore was stoped, but it is doubtful whether the mining and metallurgical operations were ever on a paying basis.

The gold deposits of the Morning Star and Pittsburg mines were opened soon after the silver deposits at Lewis. According to reports of the Director of the Mint both of these mines were being developed in 1882. The Pittsburg mine is located near the summit of a ridge east of Lewis Canyon, and in the eighties a mill was built in the canyon east of this ridge and connected by a tramway with the mine. This mill reported a production of $46,000 in 1887 and was in successful operation until 1891. The Morning Star mine, on the same ridge and southwest of the Pittsburg, was sold in 1892 to W. E. Dean, of San Francisco, who built a 10-stamp amalgamating mill in Lewis Canyon and ran an adit from the mill to the lode. This mine was worked for many years under the management of D. J. Bousfield. After tedious litigation with the owners of the Pittsburg mine that property was acquired in 1904 by the owners of the Morning Star and the two mines were consolidated as the Cumberland mines. The Morning Star mill has not been in operation since 1906 and the company is devoting its attention to a long adit which it is driving to develop the lodes below the present workings.

GEOLOGY AND ORE DEPOSITS.

The rocks in the vicinity of Lewis and Dean are in the main quartzites, limestones, and shales, which, on lithologic grounds, were termed Carboniferous by the geologists of the Fortieth Parallel Survey. The quartzite beds are more extensive than the other sedimentary rocks, although some of the limestone and shale members have a thickness of several hundred feet. High on the north slope of Shoshone Peak there are some rather coarse conglomerates, which presumably are the Weber. The sedimentary rocks have a general northeastward dip, but locally dip in other directions. They are intruded by grano-diorite porphyry, the somewhat altered specimens of which are grayish-green rocks composed of a dense greenish paste carrying phenocrysts of feldspar, quartz, hornblende, and biotite. Under the microscope the groundmass is seen to be a microcrystalline aggre-gate of colorless minerals and the phenocrysts are oligoclase, andesine, and quartz, with hornblende and biotite altered to chlorite, sericite, and other minerals. This porphyry may have the composition of the rock from Shoshone Peak analyzed by R. W. Woodward and described on page 25. The sedimentary rocks are also cut by quartz porphyries, some of which are glassy devitrified rocks which were formed presumably very near the surface. That below the Pitts-burg mine has the dense pasty appearance of rhyolite.

The ore deposits are fissure veins of pyritic gold ore in quartzite and in granodiorite porphyry and replacement deposits of siliceous silver ore in limestone. The gold deposits are the Morning Star and Pittsburg lodes and the silver deposits include the lodes of the Eagle, Starr Grove, Betty O'Neil, and other mines which have long since caved.

MINE DESCRIPTIONS.

Morning Star and Pittsburg mines.—The Morning Star and Pitts-burg mines of the Cumberland group are developed on different lodes. They are close together and were worked through the same adit. The Morning Star with connecting adits has over 2 miles of underground workings, which have a vertical range in elevation of 925 feet. The Concert and Mayo inclines are sunk on the Morning Star lode about 300 feet apart to the level 230 feet below the collar of the Concert incline, and from this level a crosscut adit connects with the surface on the east side of the ridge. A second adit is driven from the mill northeastward for 1,500 feet to the lode, which it inter-sects on the 460-foot level. A vertical winze, 1,335 feet from the portal of this adit, is sunk 465 feet to connect with the lowest adit, which is driven eastward for 3,500 feet to the winze. An incline, driven on the lode near the foot of the Mayo incline, connects the 460-foot level with the 230-foot level, and three levels are turned between.

The country rock is quartzite and granodiorite porphyry. The quartzite is of medium grain, gray or buff in color, and includes some thin-bedded siliceous shales. The granodiorite porphyry is a mass of large areal outcrop which intrudes the quartzite and sends off thin dikelike apophyses into it. The lode is a fissure vein which cuts both the quartzite and the porphyry and is mineralized in both. The country rock is greatly altered at some places for 50 feet from the vein. The porphyry has suffered greatest alteration and is changed to a mass of sericite and pyrite with a little calcite. Farther away chlorite and calcite are formed as an alteration of hornblende, biotite, and other minerals. The quartzite also is altered by the vein-forming solutions, the change being in the main the development of pyrite and sericite. Small amounts of galena and chalcopyrite are intergrown with the pyrite of the porphyry.

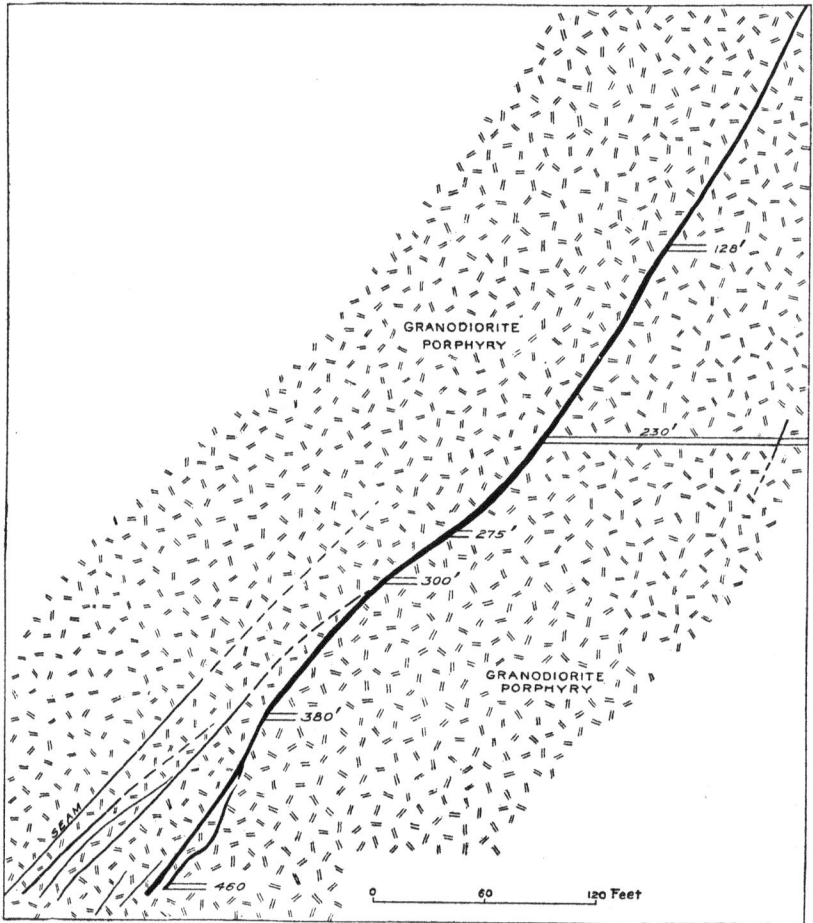

FIGURE 22.—Cross section along Mayo incline, Morning Star lode, Dean.

The lode outcrops at the surface on the east slope of the ridge and the main fissure was followed downward to the 380-foot level. It splits below this level and both branches carry good ore. On the 460-foot level there are six veins approximately parallel and in places three of them carry ore shoots. On this level the lode is composed of branching and forking, anastomosing fissures, the whole series being confined within a relatively narrow zone which strikes north and dips about 60° W. A vertical cross section of the lode in the plane of the Mayo incline is shown by figure 22. The ore is chiefly quartz, pyrite, and arsenopyrite, all of which carry free gold. Calcite and barite are present in subordinate quantities in the gangue and a little galena and zinc blende are associated with the pyrite. The quartz and sulphides are locally banded. In places comb structure is shown and drusy cavities with quartz crystals pointing to the center represent the filling of open spaces. At some places the wall rock near the ore carried working values, but the filled portion of the vein is much the most important part. Near the surface at the outcrop of the deposit, but not below a depth of 100 feet, the ore is oxidized to a brownish-yellow mass carrying considerable iron oxide and some lead carbonate. A little native copper was found on the 128-foot level in the oxidized ore, and a small amount of ruby silver is reported from the 380-foot level. The veins vary from the thickness of a knife-edge to 10 feet. The principal ore shoot which was opened between the Concert and Mayo inclines had a maximum length along the level of about 300 feet altogether and pitched approximately down the dip of the vein. It split between the 380-foot level, forming an inverted Y with ore on both legs.

The waters of the mine are acid, and this condition, together with the somewhat crushed state of the ore, would favor secondary enrichment, but little is evident from the exploration done. The sulphide ore on the 430-foot level is said to be as rich as the ore higher up and of about the same value as the oxidized ore, which did not extend more than 100 feet below the surface.

The Pittsburg mine, which joins the Morning Star on the northeast, is opened on seven levels, which have a vertical range of 550 feet and include about 4,500 feet of horizontal workings, principally drifts on the Pittsburg vein. The vein strikes west, dips 60° S., and is composed of quartz, calcite, pyrite, and arsenopyrite, with a little galena, blende, chalcopyrite, and gray copper. The country rock is quartzite and granodiorite porphyry, the latter predominating. The wall rock is altered by hot waters, as in the Morning Star. The ore carries free gold which is mainly associated with quartz and pyrite, and a small amount of silver is present, with less than 1 per cent of copper.

The principal ore shoot pitches westward in the vein at an angle of about 45° and has been stoped from near the grass roots to levels about 400 feet vertically below the surface. On the levels this ore shoot is from 50 to 150 feet long and from 2 to 10 feet wide. Another stope to the east of this ore body lies between the fifth and sixth levels and extends downward to level 7, having a vertical range of 175 feet and a length on the levels of about 140 feet.

As the Morning Star lode strikes about N. 30° W. and the Pittsburg lode strikes west, the two should intersect in the Morning Star workings on the adit level 460 feet below the Concert incline. Some small fissures which have about the same attitude as the Pittsburg vein have been exposed in the east end of the Morning Star workings, but it can not be shown now whether they represent the Pittsburg lode or sheeting parallel to it.

Starr Grove mine.—The Starr Grove mine is one-eighth mile southwest of the compressor plant of the Cumberland mines, which is one-fourth mile below the portal of the lowest adit of the Morning Star. A crosscut tunnel is driven for 150 feet to an ore body which is followed through irregular drifts and crosscuts, some 200 feet of which are still open. A deep shaft and other workings below the tunnel are inaccessible. The country rock is dark-gray limestone, which above the mine is overlain by massive quartzite beds. The deposit is a large flat body which lies approximately parallel with the bedding of the country rock, but locally cuts across it. Where developed it strikes southward and dips at a very low angle toward the west. The ore body is composed of barite and quartz, carrying here and there a small amount of pyrite, galena, and zinc blende, finely disseminated in the white gangue. At some places there is as much as 10 feet of nearly pure barite. The quartz and sulphides are in part of later origin than the barite, for at some places the crushed fragments of barite are surrounded by the quartzose ore and cracks of the ore cross the barite. On the dump some quartzose ore carrying ruby silver was found.

Betty O'Neil mine.—The Betty O'Neil mine, about half a mile west of Lewis, was worked through a 225-foot shaft which was under water when the mine was visited. A shallow adit is driven for 70 feet southwestward to the vein, which it follows for 300 feet. The vein strikes S. 30° E. and dips from 20° to 50° E. The country rock is quartzite with shaly black siliceous beds, and fragments of porphyry were noted on the dump. The vein is from a few inches to 3 feet wide and is composed of banded quartz and sulphides, chiefly pyrite, galena, and zinc blende. In 1908 a crosscut tunnel was being driven to drain the old workings.

INDEX.

O

DEPARTMENT OF THE INTERIOR

UNITED STATES GEOLOGICAL SURVEY

GEORGE OTIS SMITH, Director

BULLETIN 414

NOTES ON SOME MINING DISTRICTS IN HUMBOLDT COUNTY, NEVADA

BY

FREDERICK LESLIE RANSOME

WASHINGTON

GOVERNMENT PRINTING OFFICE

1909

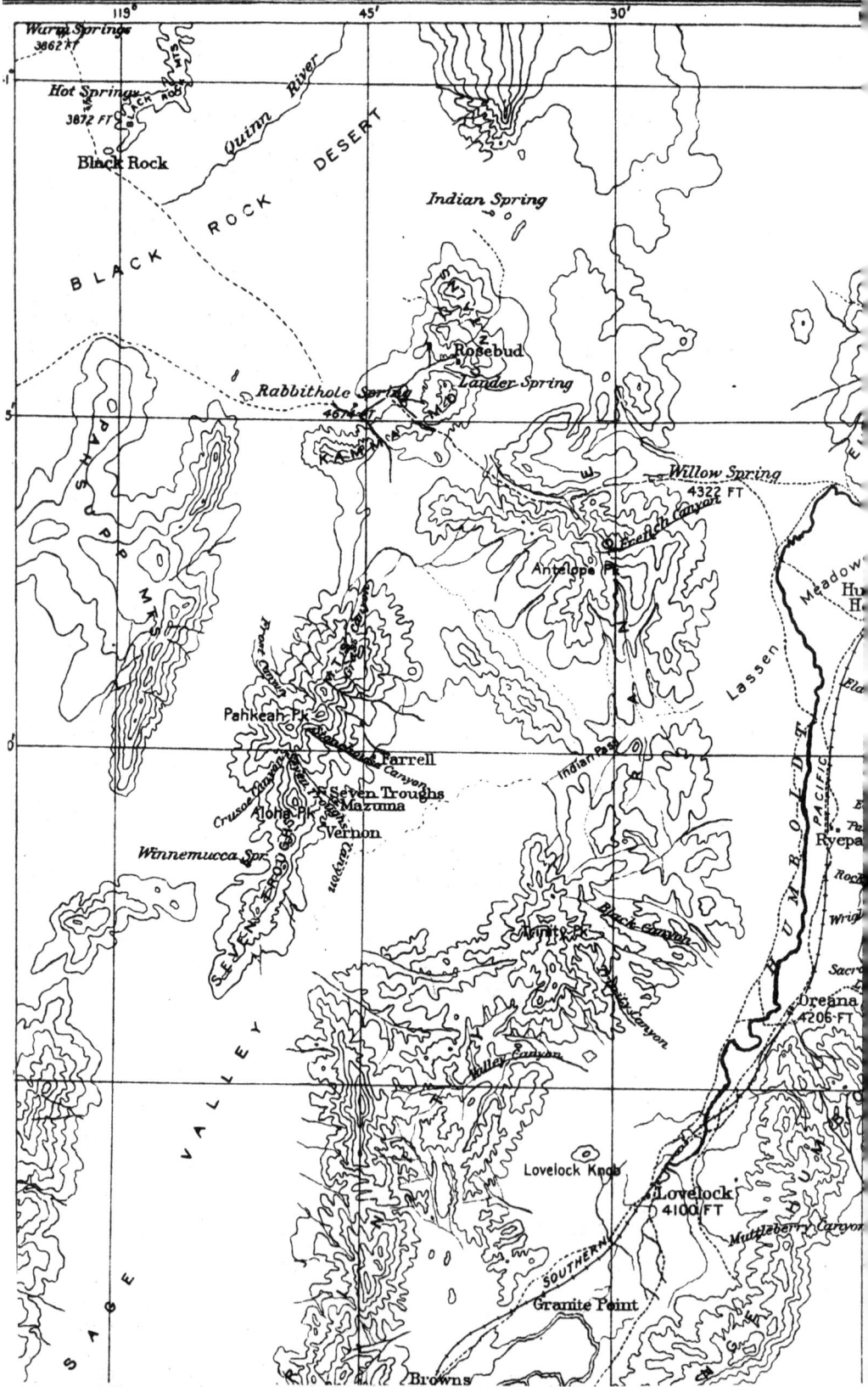

119°   45'   30'

Warm Springs
3862 FT

Hot Springs
3872 FT

Black Rock

Quinn River

BLACK ROCK DESERT

Indian Spring

Rosebud

Lander Spring

Rabbithole Spring
4671 FT

KAMMA

Willow Spring
4322 FT

Rosebud Canyon

Antelope Pk

Lassen

Meadow

Hu
He

Ela

PAH-SUPP MOUNTAINS

Pahkeah Pk

Farrell

Seven Troughs
Mazuma

Alphe Pk

Vernon

Winnemucca Spr.

Crusoe Canyon

SEVEN TROUGHS RANGE

Indian Pass

SAGE VALLEY

Arinto Pk

Black Canyon

Juniper Canyon

Valley Canyon

Lovelock Knob

Lovelock
4100 FT

Granite Point

Browns

Ryepa

Rock

Wrig

Sacr

Oreana
4206 FT

Mulberry Canyon

SOUTHERN

UMBOLD

PACIFIC

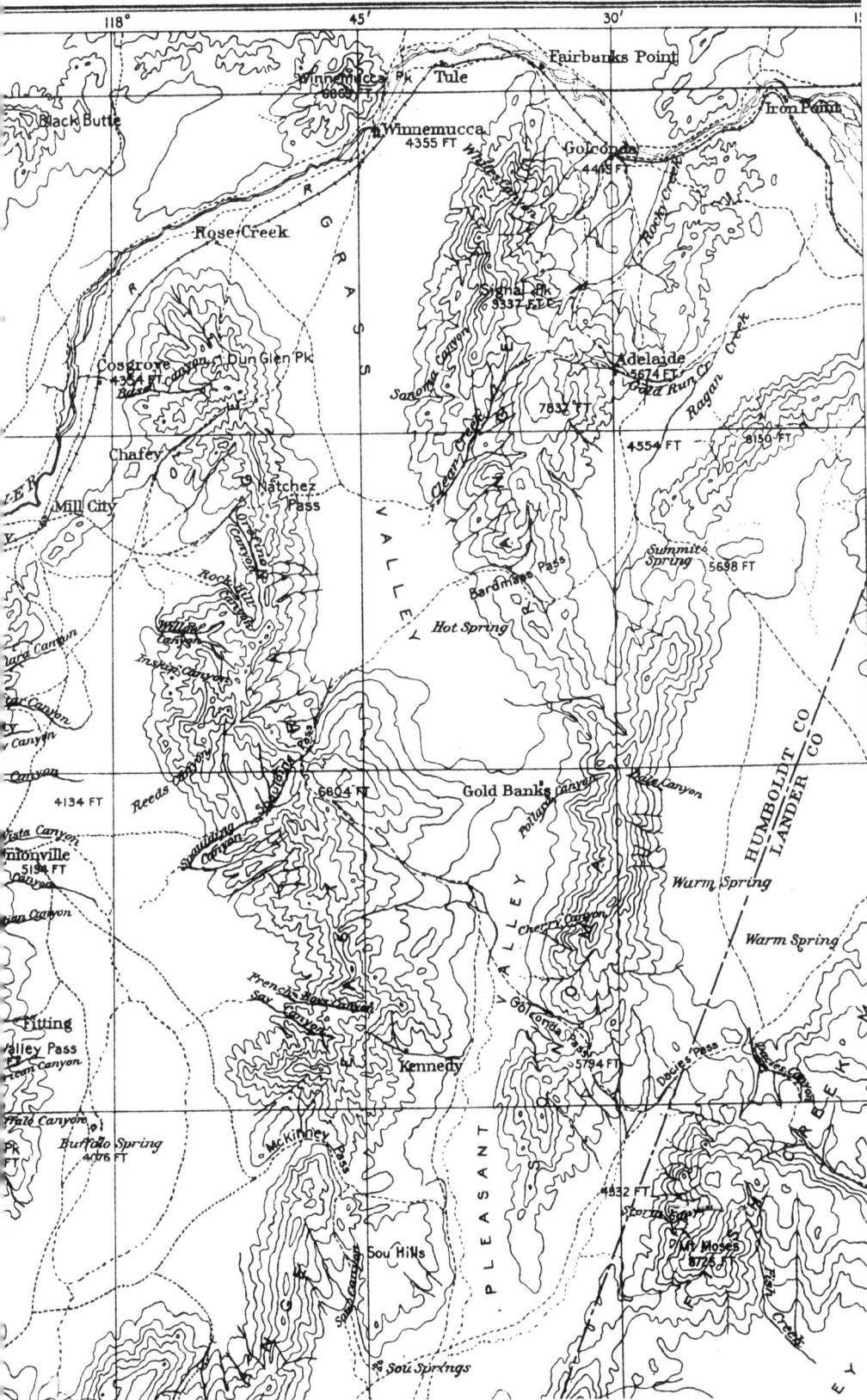

118° 45' 30' 15'

Black Butte

Winnemucca Pk
8680 FT
Tule
Fairbanks Point
Iron Point

Winnemucca
4355 FT
Golconda
4449 FT

Rose Creek

G R A S S

Signal Pk
9337 FT

Rock Creek

Cosgrove
4354 FT
Edna Canyon
Dun Glen Pk

Sanoma Canyon

Adelaide
5674 FT
Gold Run Cr.

Ragan Creek

Chafey
Natchez
Pass
Clear Creek
7832 FT

4554 FT
8150 FT

Mill City

Rock Hill

China Canyon

Bardmass Pass
Hot Spring
Summit Spring
5698 FT

Willow Canyon

Insper Canyon

V A L L E Y

lard Canyon

ur Canyon

Canyon

Reeds
Spaulding Canyon

6804 FT
Gold Banks
Pollard Canyon
Hole Canyon

HUMBOLDT CO
LANDER CO

4134 FT

Vista Canyon
nionville
5194 FT
Canyon

P L E A S A N T

Wurm Spring

Warm Spring

gan Canyon

Cherry Canyon

French Boy Canyon
Say Canyon

Fitting
alley Pass
ican Canyon

Kennedy

Golconda Summit
5794 FT

Dixie Pass

eld Canyon
Pk
FT

Buffalo Spring
4076 FT

McKinney

4832 FT
Storia Spring

Mt Moses
8725 FT

Sou Hills

Sou Canyon

Sou Springs

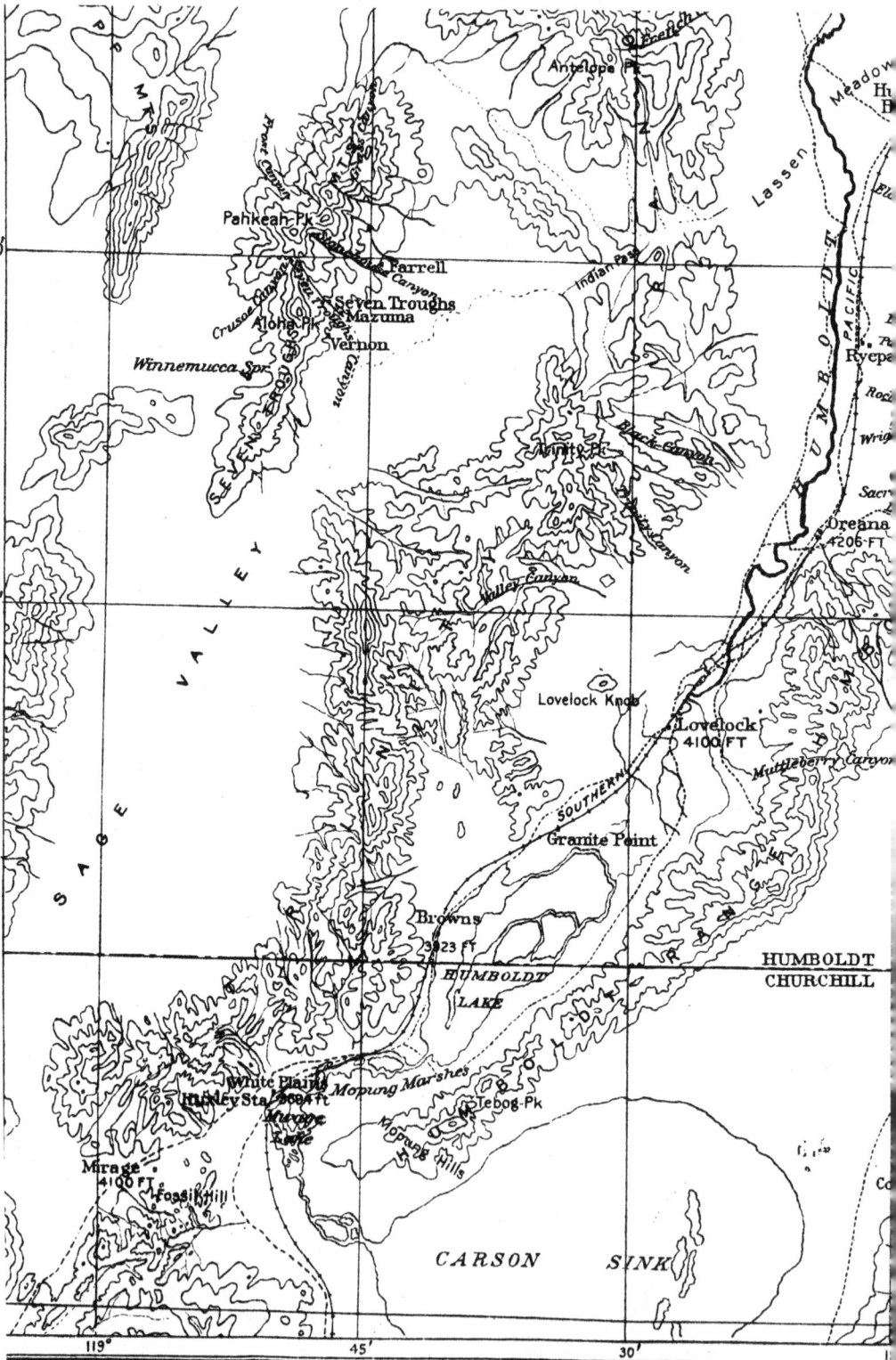

Po
Ma
 range

Antelope Pk

Lassen   Meadow

Pahkeah Pk   Indian Pk

Farrell
Seven Troughs
Aloha Pk   Mazuma
Vernon

Winnemucca Spr.

Arinte Pk

Black Canyon

UMBOLDT

PACIFIC

Ryepa

Roc

Wrig

Sacr

Oreana
4206 FT

Valley Canyon

SAGE   VALLEY

Lovelock Knob

Lovelock
4100 FT

SOUTHERN

Mutleberry Canyon

Granite Point

Browns
3923 FT

HUMBOLDT
CHURCHILL

HUMBOLDT
LAKE

White Plains
Huxley Sta   3984 FT
Mirage
Lake

Mopung Marshes

Tebog Pk

Mopung Hills

Mirage
4100 FT
Fossil Hill

CARSON   SINK

Co

119°   45'   30'

OUTLINE MAP OF PAR

Imlay

Rock Creek

Willow Canyon

Santa Clara Canyon

Inskip Canyon

Star Canyon

Spring City

Snoddy Canyon

Canyon Canyon

Reeds Canyon

4134 FT

5804 FT

Spaulding Canyon

Buena Vista Canyon

Unionville

5194 FT

Canyon

Rosebud Canyon

Indian Canyon

Fitting

Spring Valley Pass

625 FT

American Canyon

Buffalo Canyon

Buffalo Pk.

8387 FT

Buffalo Spring

4076 FT

French Boyd Canyon

Say Canyon

Kennedy

Mt. Kinney

Sou Canyon

Sou Hills

Sou Springs

Table Mountain

Chataya Pk.

7766 FT

Boyer's Ranch

Cottonwood C.

Spring

Salt Field

Hot Springs

DIXIE VALLEY

Antimony

Shoshone Springs

ANTELOPE VALLEY

Boundary Pk.

Cone Spring

LONE HILL VALLEY

Mt. Moses

8775

Storm Canyon

4832 FT

Dacies Pass

5794 FT

Gol Conda Spg.

Chet C. Canyon

Warm Spring

Warm Spring

HUMBOLDT CO
LANDER CO

Gold Banks Canyon

Poligon Canyon

Dale Canyon

PLEASANT VALLEY

CR E EK VALLEY

Hot Spring

Bardmass Pass

Summit Spring

5698 FT

Fish Creek

118°   45'   30'

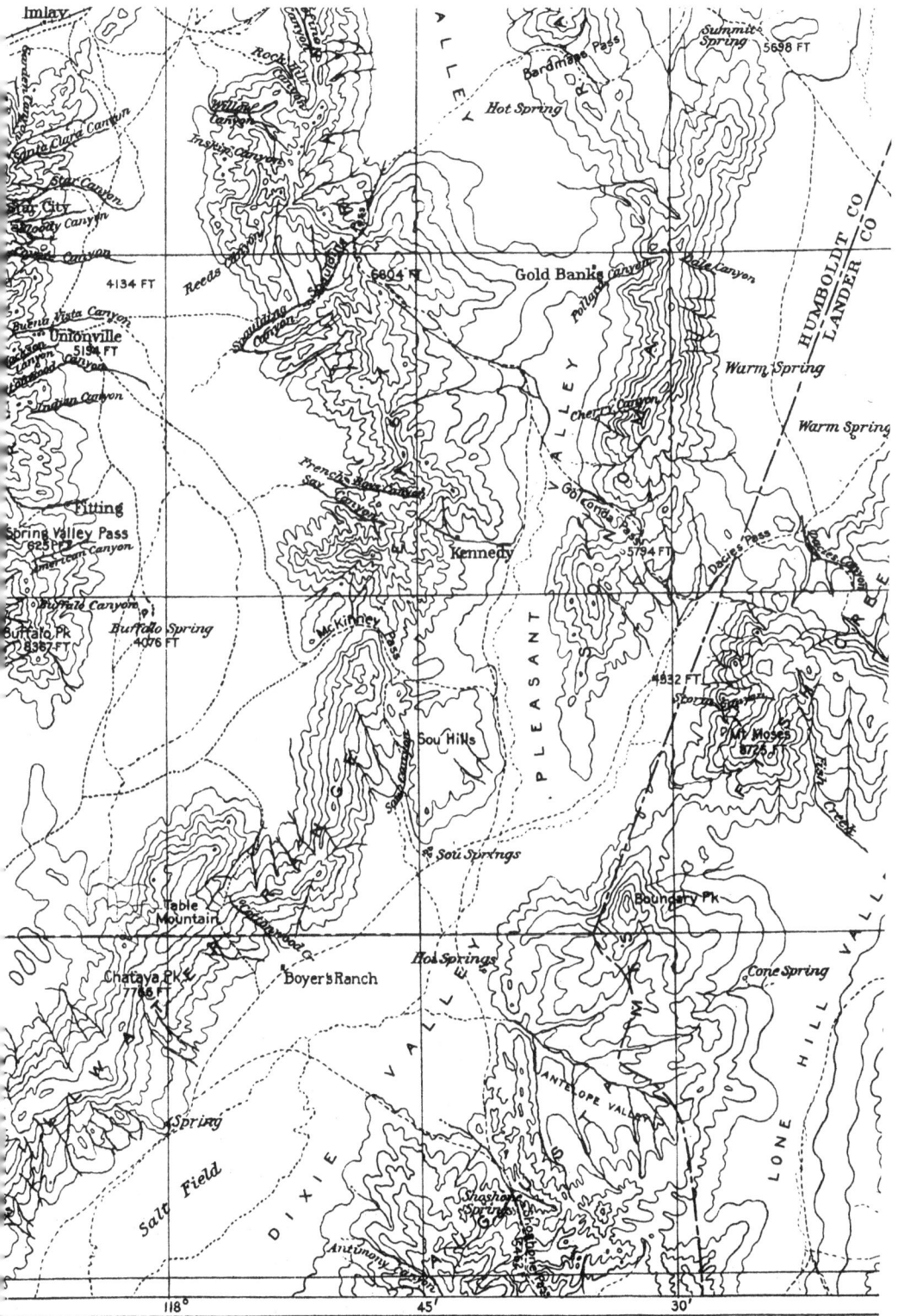

HUMBOLDT, LANDER, AND CHURCHILL COUNTIES, NEVADA

0   5   10   15 Miles

ILLUSTRATIONS.

5

CONTENTS.

CONTENTS.

NOTES ON SOME MINING DISTRICTS IN HUMBOLDT COUNTY, NEVADA.

By Frederick Leslie Ransome.

INTRODUCTION.

In the apportionment of my field season in 1908, a period of six weeks was allotted to a reconnaissance examination of that part of Humboldt County, Nev., lying between the fortieth and forty-first parallels and the one hundred and seventeenth and one hundred and nineteenth meridians. Within this area of about 7,000 square miles (see fig. 1) are the Seven Troughs, Rosebud, Star, Unionville (Buena Vista), Humboldt, Fitting, Chafey (Dun Glen, Sierra), Kennedy, and Adelaide (Gold Run) districts,[a] with many others of less note. North of the region particularly investigated is the Red Butte district and south of it is Coppereid (White Cloud district). Both of these were visited.

Some of the districts examined, such as Seven Troughs, Rosebud, and Red Butte, have been prospected only within the last two or three years; others, like Star City and Unionville, reached their acme of productiveness in the decade beginning with the year 1860 and have not yet participated in the recent general revival of mining activity in Nevada; one or two, like Chafey, have received new names and are being exploited in various ways that modern experience and ingenuity have devised for this purpose; still others, like Adelaide, have been intermittently active for over thirty years, oscillating between prosperity and decay.

All of the country traversed, with the exception of that adjacent to Red Butte, was mapped geologically by the Fortieth Parallel Survey, and that map,[b] on a scale of 4 miles to the inch, with a contour interval of 300 feet, is still the best one, although there have been many

---

[a] In general the names of districts as here used are those of the principal settlements. Some of the mining districts, as originally organized for purposes of record and regulation, embrace large tracts of which the names and boundaries have little significance except to local surveyors and official recorders.

[b] U. S. Geol. Expl. 40th Par., atlas, Map V, 1876. (The topographic sheets are not contoured, but show the relief by shading.)

changes in place names in the thirty-three years that have elapsed since its publication. In general, the "granites" of this region, described as Archean in the Fortieth Parallel Survey reports, are intrusive in

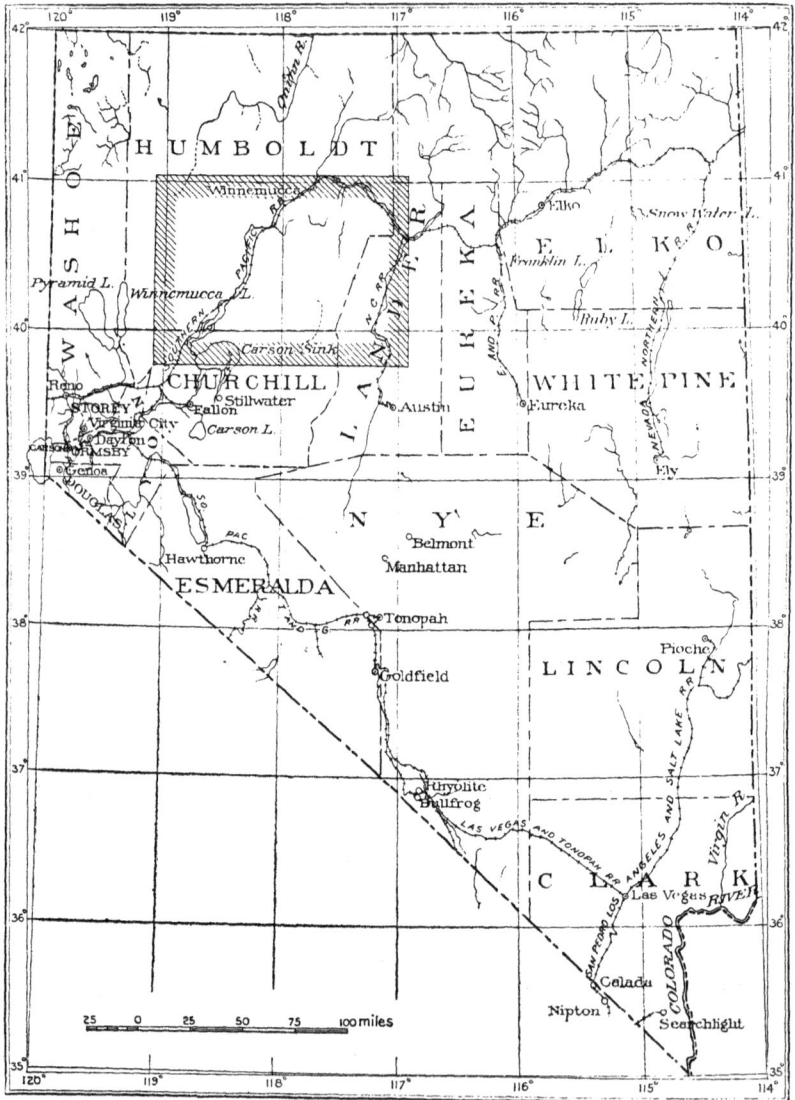

FIGURE 1.—Index map of Nevada, showing the area covered in part by the reconnaissance of 1908.

Mesozoic rocks; much of the material described as quartzite in the Triassic is rhyolite, and thorough study would change many of the names applied to the igneous rocks. During the last thirty years the mining districts of Humboldt County have received little attention from geologists, and the present reconnaissance was undertaken in the belief that even a hasty review of the region, while not likely to yield results of much scientific importance, might serve as the basis for a preliminary report that should be of some value to those interested in the mining development of north-central Nevada.

It is a pleasure to acknowledge my indebtedness to the mining men of the region for courtesies too numerous to mention, and especially to Mr. John T. Reid, of Lovelock, and Mr. W. D. Adamson, of Winnemucca, who gave generously of their time and information.

ITINERARY.

From Lovelock, on the Southern Pacific Railroad (see Pl. I), a trip of five days' duration was made to the Seven Troughs district, 30 miles northwest of that town, and return. Two days were next spent in visiting Coppereid, in the White Cloud district, Churchill County, about 25 miles southeast of Lovelock. A wagon and team were then hired at Lovelock, and the Humboldt Range was crossed by way of the Humboldt Queen mine and Limerick and American canyons to Fitting, or Spring Valley, as it is sometimes called. Thence I drove north to Unionville, and from that place turned southeast across the East Range to Kennedy, on the west side of Pleasant Valley. From Kennedy the route was south, past Sou Springs to Boyer's ranch, on the northwest side of Dixie Valley, which was a convenient place from which to examine the nickel and cobalt mines of Cottonwood Canyon, in the Stillwater Range. From Boyer's ranch the return to Lovelock was made over the Stillwater Range by way of Kitten Spring, across the valley of Carson Sink, and through Cole Canyon, east of Oreana, which separates the Humboldt Range into two distinct divisions, the northern one being sometimes referred to as the Star Peak Range and the southern one as the Humboldt Lake Range.[a] This trip occupied six days.

From Ryepatch, on the Southern Pacific Railroad, a visit was paid to the Ryepatch mine and from Humboldt House to a cinnabar prospect in Eldorado Canyon, on the west side of Star Peak. Humboldt House also was the starting point for a trip lasting three days to the Red Butte and Rosebud districts. The Chafey district and the Sheba mine in Star Canyon were reached from Mill City, and a short excursion was made from Golconda, 12 miles east of Winnemucca, to the Adelaide mine. The Galena and other districts accessible from Battle Mountain were not visited, the reconnaissance having already taken more time than could well be spared from other duties.

---

[a] Louderback, G. D., Basin range structure of the Humboldt region: Bull. Geol. Soc. America, vol. 15, 1904, p. 294.

GENERAL HISTORY OF MINING DEVELOPMENT.

Mining activity began in this part of Nevada about the year 1860, with the organization of the Humboldt district, on the northwest slope of Star Peak. At that time the Central Pacific Railroad Company was not yet incorporated, supplies and machinery were hauled tediously in wagons from Marysville or Sacramento, and the project of a transcontinental railway was little more than a dream. The Star and Buena Vista districts were organized in 1861, and during the civil war and the recovery from that conflict these isolated desert communities attained their greatest prosperity. Mines were opened along both flanks of the Humboldt Range, and the settlements of Star City and Unionville soon grew to importance. At Oreana, on the banks of Humboldt River, smelting works were built to treat ore from the Montezuma mine, in the Trinity district, organized in 1863. Special interest attaches to this smelter, as it was the first in Nevada from which lead was shipped in commercial quantities, and it contests with Argenta, Mont., the honor of being the birthplace of the present gigantic silver-lead smelting industry of the United States. Its history, however, was brief; after various failures the furnaces were at length operated successfully in 1865, but were practically abandoned about 1870. The metallurgical processes employed have been described by R. W. Raymond [a] and James D. Hague.[b] The fuel was charcoal and the products of the furnace, silver bullion and an alloy of lead and antimony, were shipped to San Francisco.

On the east side of the Humboldt Range the discovery of a rich body of silver ore in the Sheba mine close to the surface led to the rapid growth of Star City from 1861 to 1865. The town had two hotels, express and telegraph offices, daily mails, and a population estimated at about 1,000. In 1871 Raymond [c] reported the town as nearly abandoned, although at that time additional ore had been found below the levels from which the original bonanza had been stoped. At present only two men are living in the canyon, the mines are idle, two or three ruined stone cabins are all that remain of the town, and the little brook, once foul with tailings and town refuse, is now the home of trout and sparkles through diminutive hay meadows.

At Unionville the principal mine was the Arizona, owned and operated by Fall & Temple. Other important workings were those of the Silver, Pioneer, and Manitowoc mines. The Arizona mine is said to have been bought by John C. Fall in 1862 for $5,000. Soon

---

a Mineral resources of the States and Territories west of the Rocky Mountains, Washington, 1869, pp. 130–132.

b Mining industry, U. S. Geol. Expl. 40th Par., 1870, pp. 300–308.

c Statistics of mines and mining, etc., Washington, 1872, p. 208.

afterward he built the 5-stamp Pioneer mill, the machinery and materials for which were hauled by ox teams from Marysville, Cal. An aerial tramway was later constructed from the north workings of the Arizona mine to the mill in the canyon, several hundred feet below.   Other mills were subsequently built farther down the canyon, and in 1870 there were three mills of 10 stamps each in operation. Water power was used in the first mill, but it soon became necessary to supplement this with steam.   Since the principal mines were opened on different parts of a nearly horizontal vein that outcropped in an elliptical curve about a hill, litigation was inevitable, and after some controversy the Arizona and Silver companies consolidated in 1870 as the Arizona Association.   The average cost of mining and milling at that time was about $24 a ton, having not long before been lowered by the introduction of Chinese labor in the mills.   Skilled miners were paid $4 a day or $3 a day with board.   There was almost no gold in the ores, which were treated by pan amalgamation, the tailings, after standing for a time, being re-treated by the same process.   The milling ore averaged about $60 a ton, and some ore, shipped crude to San Francisco, ranged from $500 a ton upward.   One lot of 170 tons shipped in 1871 is reported by Raymond [a] to have netted $78,000, and the total ore mined in that year by the Arizona Association is given as 7,000 tons.   In 1873 the output had declined to 3,915 tons, of which 81 tons of about $330 grade was shipped to San Francisco.   The mine continued to be worked until 1880, since when little has been done with it.   It produced in all (including the output of the Henning or Wheeler mine, which yielded less than 5 per cent of the whole) about $3,000,000 from approximately 80,000 tons of ore.   In 1899 the property was brought by John Ross and Neal Carmichael, the present owners, who have not, however, resumed work on a commercial scale.

From 1860 nearly to 1880 Unionville, although perhaps rivaled or surpassed for a short time by Star City, was on the whole the most important town in the Humboldt region, and was the local supply point for many smaller communities in neighboring mining districts.

There was considerable activity during this period near Dun Glen (now known as Chafey), in the Sierra district.   The most productive mine in the seventies appears to have been the Tallulah, 2 miles northwest of Dun Glen.   Afterward the Auld Lang Syne mine became the leading one of the district.   The Monroe and Auburn mines also were active.

On the west side of the Humboldt Range mining was in progress at many places during the period when Unionville flourished.   Among

a Statistics of mines and mining, etc., Washington, 1873, p. 208.

the noted silver mines on that side of the range are the Humboldt Queen and the Ryepatch. The latter, which is reported to have produced over $1,000,000, was known before 1872 as the Butte mine. It has been idle for over twenty years.

It is to be noted that most of the mines in the Humboldt Range were opened and were worked most extensively before the completion of the Central Pacific Railroad. The great improvement in mining facilities brought about by railway communication was not sufficient to offset the diminution in tenor of the ore bodies, as they were followed below their enriched portions, and the decline in the price of silver consequent upon the demonetization of that metal.

Although the mines of the Humboldt Range have yielded far more silver than gold, placer mining was at one time an important industry, especially in American Canyon, 12 miles south of Unionville. Operations began there about 1881 and were prosecuted actively until about 1895. The placers were first worked by Americans, who are reported to have taken out gold to the value of about $1,000,000. The ground, however, soon passed into the possession of Chinese, who formed a considerable settlement in American Canyon and mined the gravels with skill and assiduity by drifting from countless narrow shafts ranging from 40 to 85 feet deep. How much gold they obtained is unknown, but some estimates, doubtless much exaggerated, place the total at about $10,000,000.

In Cottonwood Canyon, on the east slope of the Stillwater Range, near latitude 40°, are nickel and cobalt mines, which were opened about 1882, a car of nickel ore being shipped in that year to Camden, N. J. The Nickel mine, owned by the American Nickel Company, was worked until about 1890. It was again opened in 1904, but has been idle since 1907. A small smelter was built and a little matte, probably not over 50 tons, was produced. Attempts were made also to extract nickel salts with sulphuric acid, with what success is not known.

The Lovelock mine, a little farther up the canyon, has probably shipped about 500 tons of nickel-cobalt ore, which was hauled to Lovelock by teams returning from the Bernice silver and antimony district in the Augusta Mountains. A diminutive furnace was erected but was not successful.

The Kennedy district, also on the east side of the Stillwater Range and about 25 miles southeast of Unionville, first attracted attention in 1890. Kennedy soon became a flourishing town; mills were built and considerable work was done in the Gold Note, Imperial, and other mines. After the exhaustion of the oxidized pay shoots the amalgamating mills proved unfit for coping with the complex gold-silver-lead ores, and since 1904 the district has sunk into decay. No mining was in progress in 1908. The total output of the district probably does not exceed $120,000, mostly from ore of shipping grade.

The principal events of the past few years in the Humboldt region have been the opening of the Seven Troughs district in 1907 and the revival of mining in the vicinity of Dun Glen or Chafey in 1908. The Rosebud bubble, which collapsed in 1907, has merely added one more to a long list of such failures in Nevada. The various good and evil attributes on which mining "booms" depend are as a rule curiously blended or contrasted. Energy, hope, cupidity, credulity, and many other qualities all contribute to the local sentiment that applauds the "booster," no matter how extravagant his claims, and that defers the recognition of truth until the moment of disaster. The fact that a mining district is injured, not helped, by misrepresentation is forgotten by those who buy and stay, ignored by those who sell and go.

LITERATURE.

The following are the principal contributions to the geologic and mining literature on the Humboldt region:

EMMONS, S. F. From Reese River to Osabb Valley. U. S. Geol. Expl. 40th Par., vol. 2, 1877, pp. 636–659. (Describes the general geology.)

GABB, W. M. Paleontology, vol. 1, Geol. Survey of California, 1864, pp. 19–35. (Describes Triassic fossils from the Humboldt Range and from near Dun Glen.)

HAGUE, ARNOLD. Fish Creek and Battle Mountains (pp. 660–672); Havillah and Pah-Ute ranges (pp. 673–712); West Humboldt region (pp. 713–750); Montezuma Range and Kawsoh Mountains (pp. 751–774). U. S. Geol. Expl. 40th Par., vol. 2, 1877. (Describes the general geology.)

HAGUE, ARNOLD, and EMMONS, S. F. Region of the mud lakes. U. S. Geol. Expl. 40th Par., vol. 2, 1877, pp. 775–800. (Describes geology.)

HAGUE, JAMES D. Mining and milling in western Nevada. U. S. Geol. Expl. 40th Par., vol. 3, Mining industry, 1870, pp. 296–319. (Describes the Montezuma mine in the Trinity district, the smelting works at Oreana, and the mines of the Buena Vista, Star, Sierra, Gold Run, and Battle Mountain districts.)

HYATT, ALPHEUS, and SMITH, JAMES PERRIN. Triassic cephalopod genera of America. Prof. Paper U. S. Geol. Survey No. 40, 1905, pp. 21–23, 26, Plates XXII–XXV. (Describes the Middle and Upper Triassic faunas of the Humboldt Range.)

KING, CLARENCE. Systematic geology, with atlas. U. S. Geol. Expl. 40th Par., Washington, 1878. (A discussion of the structure and geologic history of the whole region adjacent to the fortieth parallel from the Rocky Mountains to the Sierra Nevada.)

LOUDERBACK, GEORGE DAVIS. Basin Range structure of the Humboldt region. Bull. Geol. Soc. America, vol. 15, 1904, pp. 289–346. (A full and valuable discussion of the general geology with special reference to the structure of the Humboldt Range.)

MEEK, F. B. Paleontology. U. S. Geol. Expl. 40th Par., vol. 4, pt. 1, 1877, pp. 99–129, Plates X and XI. (Describes the fossil fauna of the Humboldt Range as Upper Triassic.)

MERRIAM, JOHN C. Triassic Ichthyosauria. Memoirs of the University of California, vol. 1, No. 1, Berkeley, Cal., 1908, pp. 18–19. (Describes fossil reptilian remains from the Humboldt Range and gives references to earlier publications on the same Middle Triassic fauna.)

RAYMOND, ROSSITER W. Mineral resources of the States and Territories west of the Rocky Mountains. Washington, 1869, pp. 117–133. (Notes on the history and development of the Battle Mountain, Black Rock, Buena Vista, Central, Eldorado, Gold Run, Humboldt, Orofino, Sacramento, Sierra, Star, Trinity, and Winnemucca districts.)

———— Statistics of mines and mining in the States and Territories west of the Rocky Mountains. Washington, 1870–1874. (Contains much information concerning the development and working of the mines during the period covered.)

RUSSELL, ISRAEL C. Geological history of Lake Lahontan. Mon. U. S. Geol. Survey, vol. 11, 1885. (Describes the great Quaternary lake that occupied the valleys of the Humboldt region.)

SPURR, J. E. Origin and structure of the basin ranges. Bull. Geol. Soc. America, vol, 12, 1901, pp. 217–270. (Discusses incidentally the structure of some of the ranges visited in the course of the present reconnaissance. Mr. Spurr's own field work, however, was south of the fortieth parallel.)

[WISKER, A. L.] Chafey, Nev. Min. and Sci. Press, Nov. 7, 1908, pp. 625–626. (A good brief sketch of the history of the district and of mining conditions therein during the summer of 1908.)

SEVEN TROUGHS DISTRICT.

INTRODUCTION.

The Seven Troughs district, which is about 30 miles northwest of Lovelock, a flourishing town on the main line of the Southern Pacific Railroad, lies on the east slope of a minor range designated on the Fortieth Parallel Survey map[a] as the Pah-tson Mountains, but now popularly known as the Seven Troughs Mountains.[b] The higher parts of the mountains are dotted with junipers and the larger ravines contain small perennial streams. Grass flourishes on some slopes and for over thirty years the region has been used as a range for sheep and cattle. The watering places maintained in connection with this pastoral occupancy have given to the new mining district its name.

The road from Lovelock runs for a few miles through the irrigated farming land of Humboldt Valley and then crosses obliquely in a northerly direction a broad and relatively low part of the Trinity Range consisting of granitic and slaty rocks partly buried under rhyolitic and basaltic flows. From the northwest base of this range the road stretches straight across the bare and nearly level expanse of the northeast arm of Sage Valley for 9 or 10 miles, to the foot of the Seven Troughs Mountains. A more unsatisfactory road material than the mixture of stones and ashy soil that floors this arid basin can hardly be imagined, and the many abandoned deep-rutted tracks show that at times no road at all is preferable to one in which the depth of the dust-filled chuck holes is a subject for anxious speculation.

---

[a] U. S. Geol. Expl. 40th Par., atlas, Map V, west half.
[b] Occasionally referred to also as the Stonehouse Range.

Supplies are hauled by teams from Lovelock. Passengers may reach the district most conveniently from the same point by the ordinary stage line or by automobiles, which meet the transcontinental trains and ply over a little better road than is used by horse-drawn vehicles.

There are four little towns in the district, three of which, Vernon, Mazuma, and Farrell, are situated at the east base of the range. Vernon, the southernmost of the three, was the chief settlement in the district early in 1908, but had lost its preeminence by August of that year, most of the activity then centering about Mazuma, 2½ miles north-northeast of Vernon, and about Seven Troughs, which is 1¼ miles west-northwest of Mazuma, higher up the same canyon. Farrell, 3 or 4 miles north of Mazuma, has at no time been as important as the other settlements.

The ravine in which are the towns of Mazuma and Seven Troughs is known as Seven Troughs Canyon. North of it in order are Wildhorse, Burnt, and Stonehouse canyons, the latter embouching at Farrell. Victor Canyon is a short minor ravine between Burnt and Stonehouse canyons. All three of the main ravines contain water the year round, part of that from Burnt Canyon being piped to Mazuma.

It is difficult to determine when prospecting began in this district, but there is little record of any work prior to 1905, and it was not until early in 1907 that the veins began to attract other than local attention. The Mazuma Hills mine was opened in that year, and its success, with the subsequent discoveries in the Fairview and Kindergarten mines and in various leases, soon made Seven Troughs a familiar name in the mining press.

GENERAL GEOLOGY.

TOPOGRAPHY.

The Seven Troughs Mountains have a length of 24 miles and trend 20° east of north. The greatest width of the range is 8 miles. Its crest culminates in a group of three summits, which attain altitudes of about 3,000 feet above the desert plains that surround the range on all sides but the north. The highest and middle summit is Granite Peak. The one southwest of it, of schist, is designated Pahkeah Peak on the Fortieth Parallel Survey map. A low, broad pass, occupied according to this map by Miocene lake beds belonging to the Truckee formation, separates the north end of the Seven Troughs Range from the longer Trinity Range and from the Kamma Mountains, a small group within which is the Rosebud mining district and the Rabbit Hole sulphur mine.

EARLY EXPLORATION.

The Fortieth Parallel Survey map represents the Seven Troughs Mountains as consisting of a mass of Archean rocks and granite, exposed on the higher summits and over much of the west slope, but covered in the southern and eastern parts of the range by Tertiary rhyolite and subordinate overlying flows of basalt. An area of Jurassic rocks, about 5 miles long and 1 mile wide, is also shown on the east slope, just north of what is now the central part of the Seven Troughs mining district. Hague and Emmons [a] state that these rocks comprise "blue limestone and shales, which, on grounds of general probability, but without paleontological or distinct stratigraphical evidence, have been referred to the Jurassic formation."

The Archean rocks are described by the same writers [b] as very fine grained micaceous schists, distinctly bedded and standing at high angles. These are said to be cut by the granite, which, as Hague and Emmons point out, is of basic character, very different from the known Archean granites of the region, but closely resembling the rocks of the great intrusive masses, supposedly of post-Jurassic age, of the Sierra Nevada and of some of the prominent ranges of the western part of the Great Basin. They note also some small exposures of an older granite, probably Archean, on the west slope of the range.

The volcanic rocks are briefly described by the same explorers,[c] and some of the perlitic varieties were studied and figured by Zirkel.[d] Little information is given concerning the succession and structure of the lavas, and the localities particularly described are all outside of the district now being prospected.

Apparently no mining whatever was in progress in these mountains when Hague and Emmons visited them.

PRE-TERTIARY ROCKS.

The known ore deposits of the Seven Troughs Range are all in the Tertiary volcanic rocks, and consequently a small part only of the four days spent in the district could be devoted to the older formations. North of Mazuma the so-called Jurassic rocks appear first in Burnt Canyon as a northeast-southwest belt of indurated clay slate about half a mile wide. This belt appears to correspond at this place to a low ridge, which, after having been completely buried under lava flows, has been uncovered again by erosion. The slate extends northeast and forms the rounded foothills just west of Farrell. It was found to be exposed westward along Stonehouse Canyon nearly to its head, and the belt is thus two or three times as wide as is shown by

a U. S. Geol. Expl. 40th Par., vol. 2, 1877, p. 782.
b Op. cit., p. 776.
c Op. cit., pp. 770-786.
d U. S. Geol. Expl. 40th Par., vol. 6, 1876, pp. 208-209, Pls. VII, 2; IX, 2 and 3; and XII, 2.

Hague and Emmons, who evidently did not traverse this ravine and, naturally enough, mapped only the rhyolitic rocks visible on the ridges. At the old stone cabin from which the canyon gets its name, a mile or two west of Farrell, the slate is well exposed and dips about 40° E., bedding and cleavage being parallel. Farther up the canyon the dip varies and the slate is cut by dikes of a light-colored rock, presumably rhyolite. Near the head of the ravine the slate is intruded and metamorphosed by the "younger granite," described by Hague and Emmons. This metamorphosed material was mapped as Archean by these observers, and since the main mass of supposed Archean west of Pahkeah Peak consists, according to their description, of bedded slaty schists, there is a suggestion that these also may be metamorphosed post-Archean sediments. There was no opportunity in 1908 to investigate this problem; in fact, not having the Fortieth Parallel Survey map at hand, I was unaware, while in the field, that it had arisen.

The intrusive granitic rock at the head of Stonehouse Canyon is a fresh, medium-granular, rather dark gray rock, which evidently is not very quartzose and contains a large proportion of plagioclase. Only one specimen was collected and the mass undoubtedly contains other varieties than the one here described. The microscope shows the rock to be a granodiorite, of which the constituents are a plagioclase near andesine, orthoclase, microcline, quartz, hornblende, biotite, apatite, and magnetite. The plagioclase is a little more abundant than the alkalic fedspars.

The rock from the summit of Granite Peak was analyzed by Prof. Thomas M. Drown as follows:

Chemical analysis of granodiorite from Granite Peak.[a]

SiO_2	64.02
Al_2O_3	17.60
FeO	4.03
MgO	1.27
CaO	4.38
Na_2O	4.79
K_2O	2.62
Loss on ignition	.80
MnO	.16
	99.67

This specimen is said by Hague and Emmons to be typical of the mass, and their suggestion of its similarity to the granodiorite of the Sierra Nevada has been fully borne out by later work in that range.

---

[a] U. S. Geol. Expl. 40th Par., vol. 2, 1877, p. 779.

TERTIARY VOLCANIC ROCKS.

The prevailing rock of the low, rounded hills on the edge of the Sage Valley at Mazuma and for a mile or more up Seven Troughs Canyon is a pale reddish-brown lava, much of which shows conspicuous flow lamination and a platy fracture. Some varieties are light gray, of compact lithoidal texture, and contain minute phenocrysts of feldspar and biotite; other kinds are brittle perlitic glasses. Although the rock is not visibly quartzose, its general appearance is so rhyolitic that little doubt of its siliceous character arose in the field, especially since it was mapped as rhyolite by the geologists of the Fortieth Parallel Survey. The microscope shows, however, that neither quartz nor orthoclase is present in identifiable crystals, but that phenocrysts of labradorite and biotite lie in a very glassy groundmass containing plagioclase microlites. Some ill-defined aggregates of magnetite suggest the former presence of small phenocrysts of hornblende that have undergone magmatic resorption. The rock accordingly is not a rhyolite but, pending chemical analysis, must provisionally be classed as a mica andesite, unusually poor in femic constituents. Much of it is a mica andesite vitrophyre.

Associated with the mica andesite, which appears to occur in several thin flows, is at least one flow of vesicular basalt, some of which is exposed on the slope just north of the west end of Mazuma.

The mica andesite, with a general dip to the east, is the principal rock of the ridges on both sides of the road from Mazuma westward to Seven Troughs. Here the canyon expands into a little amphitheater formed by two short lateral ravines that open north and south on the main east-west gorge. The northern and longer ravine heads in a saddle, through which passes the trail from Seven Troughs to Wild-horse Canyon. The southern ravine, which is much shorter and steeper, rises to a higher saddle, through which goes the trail to the Fairview mine and to Vernon. The length of the amphitheater is about a mile and its width about one-fourth mile. Structurally it appears to correspond to a sharp north-south anticline, along which the mica andesite has been eroded, exposing older and generally softer rocks beneath it. Consequently the andesite in general rims the basin and the other rocks occupy its lower part.

Directly under the vitrophyric andesite there is, as a rule, a flow of basalt. The thickness of this is not known but appears to vary greatly. Much of it is vesicular, and transitions from compact to vesicular varieties are in some places abrupt. Some parts of the flow are a brittle glass and must have been very quickly cooled. Under this basalt and occupying the bottom of the Seven Troughs amphitheater is a volcanic complex of rhyolite, basalt, mica andesite, tuffs, arkosic sandstones, and possibly other rocks of which the structural relations are as yet very imperfectly known and probably can be ascertained only by careful mapping and detailed microscopic work. The oldest rocks appear to be glassy tuffs, in part rhyolitic and in part basaltic, and arkosic grits. Interbedded with these is at least one flow of glassy amygdaloidal basalt. These are cut by dikes of compact basalt and of a light-colored rock which is apparently the equivalent of the mica andesite flows already described. One dike of this rock extends from Seven Troughs northward nearly to the saddle opening into Wildhorse Canyon and forms a little ridge east of the principal mines north of town. In places it is from 200 to 300 yards wide. Near Seven Troughs it contains masses of obsidian or glass, which suggest that this part of the dike must have cooled very near the surface. Farther north it has a rude columnar structure, the columns lying horizontally across the dike. It must be said that the identification of the mass as a dike rests almost wholly upon its general form and position. The possibility of its being a down-faulted strip from the mica andesite flows has not been eliminated.

The tuffs and arkosic sediments are very poorly exposed, and a few mine workings, as yet shallow, are the only places where anything can be learned of their attitude and structure. In some of these the beds are nearly horizontal; in others they are tilted up to 45°, the dip in the few openings where it could be observed according generally with the anticlinal structure of the district. In the mines changes from tuff to basalt or rhyolite are common and often perplexing. This is due partly to the fact that the basalt occurs both as flows and as irregular intrusions, partly to the local and variable character of the different formations, and partly to faulting.

Associated with the basalt and tuffaceous deposits are masses of a light-yellowish rhyolite, which in the field, prior to petrographic study, was not readily distinguished from the younger mica andesite, then thought to be also a rhyolite. It was supposed at the time of visit that the yellowish rhyolite might be a lower, slightly altered part of the series of flows now known to be andesitic. It proves, however, to be a distinct and older rock. Whether it is extrusive and rests generally on the tuffs or is intrusive into them is not yet known. It is exposed in the bottom of Seven Troughs Canyon, just west of the town, where it is apparently separated from the mica andesite by a basalt flow. The rock is compact and contains numerous small phenocrysts of quartz, orthoclase (sanidine), and biotite. The microscope shows also a few phenocrysts of plagioclase. The groundmass is partly devitrified microlitic glass.

In general it must be said that in detail the structure of the district is obscure and complex; it will require much more than a reconnaissance examination for its satisfactory interpretation.

The Fortieth Parallel Survey map shows that a considerable part of the range in the vicinity of Aloha Peak, south of Seven Troughs, is covered by basalt. Dark scarps, presumably of this rock, are visible along the crest of the range from near Vernon, and bowlders of hard vesicular basalt are abundant in the wash near that town. This flow is probably younger than any of the rocks exposed near the mines.

DISTRIBUTION AND DEVELOPMENT OF THE MINES.

The most important group of mines is at Seven Troughs. Just southeast of the town, which lies on the south bank of the arroyo running down to Mazuma, are the Kindergarten and Wihuja mines, both on the same vein. The Kindergarten mine, owned by the Seven Troughs Kindergarten Mining Company, is developed by a tunnel and by an inclined shaft 280 feet deep on the dip. A new vertical shaft, being sunk at the time of visit in August, 1908, was expected to cut the vein at a depth of about 300 feet. The Wihuja is a lease on the ground of the Seven Troughs Therien Gold Mining Company, and is opened by an inclined shaft to a vertical depth of about 212 feet. The Kindergarten and Wihuja workings are connected. Other leases on Therien ground, in operation but not productive in 1908, were the Bard and Jess (175 feet deep), the Tyler, and the Sandifer leases.

On the north side of the canyon, close to town, are the Mazuma Hills and Reagan mines, both productive. The Mazuma Hills mine is owned and operated by the Mazuma Hills Mining Company; the Reagan is a lease on a vein lying east of the Mazuma Hills vein, but within the ground of that company.

The Mazuma Hills mine is opened by a main adit 700 feet long. Winzes from this adit connect with a level 100 feet below and about 500 feet long. There is also an upper disused adit about 100 feet above the main level. The Reagan is worked through a shaft that was 165 feet deep at the time of visit. Only the 65-foot level, however, could then be examined, the bottom level being temporarily under water, pending the installation of pumps.

South of the Mazuma Hills and Reagan mines is the Sandifer lease on Therien ground. Here, in the bottom of the canyon, a shaft is being sunk in expectation of finding ore in the southern parts of the Mazuma Hills and Reagan veins.

North of the Reagan shaft, on the same fissure zone, are the Chadbourne and Bradley leases, whose shafts are respectively 135 and 165 feet deep. Neither had been productive up to August, 1908. On the hillside a short distance above and north of the Mazuma Hills tunnels is the shaft of the Hayes-Mazuma lease. This was being sunk through rhyolite at the time of visit and was not in ore.

Between the workings mentioned and the head of the ravine north of Seven Troughs are the Eclipse shaft, Providence tunnel, and various smaller unproductive openings made by lessees and prospectors. On the north side of Seven Troughs Canyon, about a quarter of a mile below the town, is the tunnel of the Seven Troughs Tomboy Mining Company. This is a crosscut running N. 50° E. At the time of visit it was 800 feet long, and the intention of the company was to carry it 400 feet farther. The tunnel first penetrates about 150 feet of rhyolite and then goes through a seam of gouge into soft pyritized tuffaceous beds with a general low dip to the northeast. These are cut by many faults, probably of small throw, and contain some masses of basalt. About 350 feet from the portal the tunnel goes through another seam of gouge into rhyolitic (or possibly andesitic) breccia cut by dikes of glass or obsidian. Lower down the canyon, near Mazuma, considerable tunneling has been done on the Badger group of claims. These workings were not examined.

In Wildhorse Canyon prospecting was in progress in 1908 on the Wild Bull, North Pole, and other claims. The Wild Bull showed a little ore, but no shipments had been made.

North of this canyon the only active prospecting appeared to be on the Snow Squall claim in Victor Canyon, south of Farrell. It was reported that lessees had found good ore in sinking their shaft, but the workings were not visited.

From the saddle south of Seven Troughs a long ravine runs south and then turns southeast, embouching at Vernon. In the upper part of this ravine is the Dixie Queen shaft 230 feet deep and the Cleghorn Consolidated and Signal tunnels from 200 to 300 feet in length. Some lessees also were operating in 1908 on property of the Signal Peak Mining Company, high up on the ridge south of Seven Troughs, but owing to lack of time their shaft was not visited. About halfway down the canyon and about 2 miles south of Seven Troughs is the Fairview mine, reported at the time of visit to be 650 feet deep. This mine is known to have had some bunches of very rich ore in the upper levels and is said to have shipped about $65,000. No stoping was in progress at the time of visit and the shaft was being carried down through hard basalt. The mine is owned by practically the same people that control the Kindergarten and Therien properties at Seven Troughs. In contrast to the attitude of other mine owners in the district they showed disinclination to impart information and refused access to the Fairview mine.

Adjoining the Fairview workings on the north is the Harris lease, on Fairview ground, with a shaft 185 feet deep. The dump is basalt, much of it being vesicular.

There were two mills in operation in the latter part of 1908, one belonging to the Kindergarten company and situated at Seven

Troughs, the other belonging to the Mazuma Hills company and situated at Mazuma. Both are 10-stamp mills with amalgamating plates, Wilfleys, and vanners. No attempt is made to cyanide the tailings or save them.

CHARACTER OF THE DEPOSITS.

Most of the dikes and fissures near Seven Troughs have a nearly north-south trend. The veins as a whole consist of soft, crushed material and do not outcrop above the surface. They represent zones of brecciation or of small fissures, along which movement has continued since the spaces originally formed were filled with quartz. Consequently the typical quartz of the district is friable or sugary, and generally contains or is mingled with many fragments of shattered rock. Clear solid masses of quartz, even of small size, are rare. The veins on the whole are rather narrow, ranging from a few inches up to about 2 feet in width. It is possible, however, that the average working width may be considerably increased when the district has better facilities for handling and treating ore.

The valuable constituent of the lodes is native gold containing a considerable proportion of silver, and consequently of a rather pale color. In most of the rich ore the gold is visible either as clusters of small irregular particles or as coarse crystalline aggregates. No complete well-formed crystals were seen, but there is a noticeable tendency of the coarser gold to form crystal facets. The Mazuma Hills, Reagan, and Fairview mines have afforded some very showy specimens of bright yellow gold interlaminated with firm quartz or enveloping fragments of altered country rock. Loose nugget-like masses up to an ounce in weight have been found in soft crushed vein matter in the Reagan lease. The rich bunches of gold are not uniformly distributed through the veins, and it is difficult in some cases to secure clean sorting. It was found, for example, that material thrown over the Reagan dump as waste or as low-grade ore to be treated later carried small quartz stringers, and that some of these, when broken across, contained coarse native gold.

The tenor of the ores, as is to be expected, has a wide range. A mill run from the Reagan lease in August, 1908, averaged about $130 a ton. Picked ore from the Fairview, Mazuma Hills, and Reagan mines has yielded at the rate of several thousand dollars a ton. In the Kindergarten mine the ratio of gold to silver by weight is said to vary from 1 : 2 to 1 : 3 near the surface, but at the bottom of the mine to be nearly 1 : 1. Assay certificates of rich ore from the Reagan lease, seen in Seven Troughs, showed a ratio of nearly 2 of gold to 1 of silver. At one place on the lower level of the Mazuma Hills mine quartz carrying a little chalcopyrite is reported to have yielded on assay 200 ounces of silver and 0.3 ounce of gold to the ton.

The three important veins near Seven Troughs are the Mazuma Hills, Reagan, and Kindergarten veins. The first two, known only on the north side of the canyon, strike about N. 10° E. and dip from 60° to 65° W. The Reagan vein lies about 40 feet east of the Mazuma Hills vein. The Kindergarten vein, on the south side of the canyon, strikes N. 63° E. and dips south. The dip varies from 60° near the surface to 22° on some parts of the 40-foot level. The dip at the bottom level is about 35° (vertical depth 212 feet). The stope length of the ore shoot is about 130 feet, and the average workable width of ore is probably about 2 feet. At the northeast end of the mine the vein appears to be cut off by a zone of north-south fissures in which no ore had been found at the time of visit.

The country rock varies from place to place in each mine. The Kindergarten and Wihuja workings, as seen in 1908, are mainly in basalt. Part of this is a soft, altered amygdaloidal variety, evidently an extrusive rock. Other parts are a dense, olivinitic variety that apparently cuts the amygdaloidal flow rock. Masses of soft, light-colored rock, either rhyolite or mica andesite, but too decomposed as a rule for satisfactory determination, occur at unexpected places on both sides of the vein. They probably represent intrusions more or less displaced by faulting.

The Mazuma Hills vein follows a basalt dike that varies in width from a fraction of an inch to more than 6 feet. The general country rock is a nearly white altered rock that the microscope shows to have been originally a highly glassy rhyolitic tuff or flow breccia. It is now devitrified into a fine-grained, obscurely crystalline aggregate and contains minute disseminated crystals of pyrite. The best ore is on the foot wall or east side of the dike, and appears to be for the most part minutely fissured and silicified rhyolite tuff; but some ore extends into the basalt. At one place there is a second vein, about 10 feet east of the main fissure zone and with a little lower dip. The horse of rhyolite tuff between the two is said to be all low-grade ore. Some of the best ore in the mine is reported to occur as small gold-bearing stringers traversing hard rhyolitic tuff.

The Reagan vein also follows a basaltic dike and shows considerable resemblance to the Mazuma Hills vein. The general country rock, however, is more varied. Much of it is a highly-altered volcanic glass, apparently basaltic, which owing to its originally brittle character, has been elaborately and minutely cracked and fissured. Many of the cracks are microscopic and have a perlitic arrangement. This rock is generally greenish gray, and much of it is so fine grained that its mineralogical character can not be ascertained under the microscope. Other varieties show a few microscopic phenocrysts of plagioclase, partly altered to calcite, and a groundmass in which traces of an original microlitic texture can be detected. Much of the rich ore of the Reagan lease consists of this altered basaltic glass, in which the irregular cracks have been filled with quartz carrying free gold. The secondary minerals identified in the glass itself are pyrite, quartz, calcite, and apparently a little chlorite. Calcite is not abundant, which is rather surprising in view of the calcic composition of basalt. In some places altered glass of the kind described passes into a highly amygdaloidal variety that is well exposed in a tunnel on the Sandifer lease just south of the Reagan mine and in the Bradley lease to the north. In the Sandifer lease the vesicles are only partly filled by clear projecting crystals of quartz. In the Bradley lease some are filled with quartz and some with calcite. Pyrite is disseminated through the light greenish gray altered substance of both varieties, and a few vesicles were noted that were first lined with pyrite and then filled with quartz. Although pyrite is fairly abundant throughout the altered basaltic and tuffaceous rocks near the ores, and occurs with quartz in very small, almost microscopic veinlets, it appears to be rare in the generally larger veinlets in which are the visible particles of gold. These veinlets, so far as could be seen in 1908, contain little or no pyrite. According to Mr. D. H. Skea,[a] some proustite and possibly some stephanite or polybasite were found with the gold in the Fairview mine, and a little chalcopyrite and specks of a gray mineral resembling bornite were noted in 1908 on the bottom level of the Mazuma Hills mine in quartz similar to that elsewhere rich in gold; but as a rule the auriferous quartz is notably free from any other mineral than gold. Stibnite occurs in friable lenticular masses of considerable size in soft crushed basalt in the Chadbourne lease, and is said to have been found also in the Reagan lease. It does not, however, appear to have any intimate connection with rich ore. A little native silver is said to have been panned from the ore of the Wild Bull mine in Wildhorse Canyon.

All of the ore visible in 1908 was within the range of oxidation. The results of weathering, however, owing to the very small quantity of pyrite in the veinlets, are not conspicuous, and there appears to be no very definite change from oxidized to sulphide ores. Pyrite and stibnite, as has been seen, both occur in connection with the ore deposits above the ground-water level. The surface of the underground water near Seven Troughs corresponds approximately to the bottom of the canyon, in which is some running water, and doubtless rises a little higher in the adjacent ridges. In August, 1908, water was just making its appearance in the bottom of the Kindergarten mine at a depth of 212 feet, and the lessees of the Reagan, 165 feet deep, were putting in a pump in order to work their lower level.

---

[a] Oral communication.

OUTLOOK FOR THE DISTRICT.

Not enough mining or geologic work has been done to enable anyone to pass final judgment on the future of the Seven Troughs district. The presence of very rich, easily treated gold-silver ore in fair abundance at several places within an area some 6 miles in length is highly encouraging. On the other hand, it should be noted that the veins are not of great size and apparently are not as a rule of great length or persistence, and their character at any considerable depth below possible superficial enrichment is yet undetermined. Moreover, it is evident that in most of the mines the country rock may be expected to differ at various depths, and it is yet to be proved that the rocks beneath the tuffs and basalt in which most of the known ore occurs will be equally productive. In short, while the district is a most promising one for prospecting and developing, it is yet too early to regard it as one certain to yield largely for a period of many years.

ROSEBUD DISTRICT.

SITUATION AND HISTORY.

The Rosebud district is situated in the Kamma Mountains, a minor crescentic ridge lying north of the Seven Troughs Mountains and fronting with its concave northwest side the forbidding expanse of the Black Rock Desert. The main summits rise from 2,000 to 3,000 feet above the desert. Like the Seven Troughs Mountains, the group is merely a part of the Trinity Range. The town of Rosebud, close to which the mines are situated, is about 28 miles northwest of Humboldt House, a station on the main line of the Southern Pacific Railroad, and about 35 miles from Mill City. There is a stage from Humboldt House about once a week, but in September, 1908, no attempt was being made to maintain a regular schedule.

The Kamma Mountains have long been known as a source of sulphur, the Rabbit Hole sulphur mine having been worked since 1874. This deposit, which has been described by G. I. Adams,[a] is about 5 miles north-northeast of Rosebud and was not visited in the course of the present reconnaissance.

Silver ore appears to have been first discovered near the site of Rosebud in 1906. This was followed by a senseless "boom," in which, as usual, folly played eagerly into the hands of fraud. Consequently, the town, which sprang up before the existence of any considerable body of ore was assured, is now nearly deserted, and the winds whistle through the unglazed windows of its most pretentious buildings, abandoned before completion.

---

a The Rabbit Hole sulphur mines near Humboldt House, Nevada: Bull. U. S. Geol. Survey No. 225, 1904, pp. 497-500.

A year or two ago the Brown Palace mine shipped from 15 to 20 tons of ore from an open cut about 10 feet deep. This, so far as is known, represents the total output of the district up to the end of 1908. Some prospecting and development were in progress in September of that year, and it is possible that with the opening of the Western Pacific Railroad, which passes a few miles north of Rosebud, the district may ultimately prove of some importance.

GENERAL GEOLOGY.

According to the Fortieth Parallel Survey map, the oldest rocks in the Kamma Mountains are Triassic and are exposed in the southwest part of the range, associated with a little dacite and diorite. About Lander Spring, on the east side of the range, is shown an area of Jurassic rocks overlain to the west by rhyolite and andesite that are probably Tertiary. The supposed Jurassic slates are the prevailing rocks seen on the road from Humboldt House to Rosebud, which crosses the Trinity Range north of Antelope Peak, by way of Willow Spring. The veins worked at Rosebud are all in the rhyolite, which occurs as laminated flows and flow breccias. These rocks form rounded hills, rising about 1,500 feet above the town. The rhyolite is not conspicuously porphyritic and as a rule is very fine grained. Most of it represents an originally glassy rock that has undergone devitrification and locally a good deal of secondary alteration—chiefly silicification, with the development of some sericite, kaolin, and pyrite.

ORE DEPOSITS.

The principal zone of mineralization lies about half a mile north of Rosebud and strikes nearly east and west. At the west end of the zone is the White Alps property, on which some lessees have an inclined shaft 130 feet deep on a belt of fissuring and alteration that dips 70° N. There is very little vein quartz, and that merely in the form of small irregular stringers. The mineralized material is a white, much-altered rhyolite containing abundant finely disseminated pyrite which here and there gathers into small bunches. The individual crystals are as a rule almost microscopic. The soft white material into which the rock is largely altered and in which the pyrite is embedded is kaolinite. The deposit appears to have no definite walls and is generally of low grade. A little $50 ore has been found near the surface.

East of the White Alps, on the same general zone of fissuring and alteration, is the Brown Palace mine. Here two tunnels, one 300 feet and the other about 500 feet long, have been run into the hill on opposite sides of a small spur. No ore has been found in them. Higher up the hill is a small open cut on the fissure zone, whence was obtained the only ore shipped from the district. This occurred as a mass of 15 to 20 tons, in which the valuable constituent was massive argentite associated with kaolin, limonite, yellow pulverulent jarosite, and more or less oxidized rhyolitic material. This ore was found at a place where a little north-south fissure with a low dip to the west joined the main fissure zone on its south side. The ore rested on the foot wall of the minor fissure and did not extend over 10 feet from the surface.

East of the Brown Palace ground is the Dreamland, with a 100-foot vertical shaft and short drifts on two levels. The vein strikes N. 70° E. and is about vertical. It consists of hard, dull-white cryptocrystalline quartz in pyritized rhyolitic flow breccia. Its maximum width is 1 foot, but it is irregular, splits at some places into stringers, and apparently is not very persistent. The quartz contains more or less disseminated argentite, and a little ore has been sorted and saved for shipment.

There are a few other prospects in the district at which work was in progress at one time or another in 1908, including the Golden Anchor east of the Dreamland. These, however, were not examined.

Although the rocks of the Rosebud district have evidently been subjected to the action of solutions similar to those that elsewhere have produced important deposits of the precious metals, it is not yet apparent that any large or persistent veins have here been formed. The little ore thus far discovered is so near the surface and is of such a character that it can not be regarded as indicating deep and important ore bodies, although the possible existence of these is not denied.

RED BUTTE DISTRICT.

SITUATION AND HISTORY.

The small settlement of Red Butte, which consisted in September, 1908, of about 30 tents, is situated 45 miles north-northwest of Humboldt House, 55 miles a little north of west from Winnemucca, and 30 miles north of Rosebud. It lies near the south end and on the west slope of a rugged ridge, locally known as the Jackson Range, but really connected by a broad belt of lower hills with the Trinity Range to the south. There is no regular stage to Red Butte, but the district may be reached by hired conveyance from Humboldt House or Mill City. The camp is attractively situated close to a flowing spring, on a gentle slope backed by the dark, partly wooded peaks of the Jackson Range and fronting westward on the vast gleaming expanse of the Black Rock Desert, broken here and there by lonely buttes whose strange and sharply carved forms glow with the ethereal colors of a desert landscape.

The Red Butte district was opened by Mr. A. D. Ramel in May, 1907, his first prospecting being for gold on the Redeemer claim about a mile north of the present settlement. Since then work has been directed mainly to the exploration of the various copper deposits presently to be described. No ore has been shipped, and underground exploration at the time of visit was confined to openings of a very superficial character made for the most part by the original locators of the claims. The quiet prospecting here in progress is very different from the methods that gave notoriety to Rosebud.

GENERAL GEOLOGY.

For over 15 miles north of Humboldt House the road keeps to the middle of a wide, deeply filled desert valley. It then turns northwestward and crosses obliquely a broad, low part of the Trinity Range composed principally of dark slates mapped as Jurassic by the geologists of the Fortieth Parallel Survey [a] from their lithologic resemblance to the Mariposa slate of California and to slates overlying Jurassic limestone in the Humboldt Range. The grade for the Western Pacific Railroad has now been carried through these hills, and the slates have been well exposed in numerous deep cuts, which, however, there was no opportunity to examine on this trip.

From these slate hills the ragged pinnacles of the Jackson Range, which appear to rise 2,000 to 3,000 feet above Black Rock Desert, are conspicuous to the north. They are composed, so far as could be judged from an examination of their lower slopes and from material brought down by streams, of a hornblendic and hypersthenic gabbro, which also is the prevailing rock of the Red Butte district. The gabbro and related igneous rocks are probably intrusive into the slates, but no exposure of the contact was seen in the few hours spent in the district. Some stratified rock, reported to be limestone, is visible from the camp in a subordinate ridge along the edge of Black Rock Desert, a few miles to the northwest.

The most abundant gabbroitic rock near Red Butte is dark gray, medium grained, and evidently rich in hornblende. The microscope shows it to consist of a hypidiomorphic-granular aggregate of labradorite, hornblende, augite, quartz, magnetite, and apatite. The hornblende appears to be in part primary, but in part also it has been derived from the augite. Most of the quartz is an original constituent, and fills angular interstices between the feldspar crystals. This rock as a rule is more or less altered and contains quartz, sericite, chlorite, and a little epidote derived from the other minerals. A more coarsely crystalline variety of lighter color was noted about 1 mile north of the camp. This contains crystals (poikilitic anhedrons)

---

[a] Described in vol. 2, 1877, pp. 715-756.

of hornblende up to 2 centimeters in length. Under the microscope the rock is seen to be composed of calcic labradorite, augite, hypersthene, biotite, quartz, and much hornblende. The last is in part sharply intergrown with the pyroxenes, but a considerable proportion of it appears to have been formed by the alteration of the colorless augite.

Detailed petrographic study would doubtless reveal other varieties of basic igneous rocks, but for the purposes of this reconnaissance the general country rock of the district is designated with sufficient accuracy as gabbro.

Cutting the gabbro are numerous dikes of a light-gray to pink aplitic rock that contains practically no dark or femic constituents. These dikes vary greatly in size and trend, and some of them are very irregular, as may be seen on Anaconda Ridge just southwest of the settlement. The contacts between the dikes and the gabbro, as exposed on this ridge, are close and distinct, but careful inspection shows that between the two rocks there is generally a gradational zone, less than an inch in width. Along some of the larger dikes the zone is wider, and this suggested at first glance that the dikes were merely belts of alteration in the gabbro. The dikes probably were injected while the gabbro mass was still at a high temperature, so that instead of a rapid chilling of the dike magma at its bounding walls there was a slight interpenetration and mixture of dike material and gabbro. This feature is not uncommon where granitic rocks are cut by the aplitic dikes characteristically associated with siliceous plutonic intrusives.

Microscopical study shows the pinkish dike rock to consist of alkalic feldspars with some sodic oligoclase and considerable interstitial quartz. In some varieties the quartz and feldspar are intergrown as micropegmatite. Most of the alkalic feldspars are twinned repeatedly and irregularly according to the albite and pericline laws, and are intergrowths of both the albite and orthoclase molecules. Specific determination of them was not attempted.

A considerable part of Anaconda Ridge, about 2 miles southwest of Red Butte camp, is composed of a compact, minutely porphyritic, dark-gray rock, which the microscope shows to be a slightly altered andesite. Presumably this rests on or is intruded by the gabbro, but its structural relations were not ascertained. Some basaltic rocks also were noted along the road a few miles south of the camp, between the gabbro and the slates, but they were not closely examined. One kind of these is remarkable for the large size of the plagioclase phenocrysts, which are tabular parallel to the brachypinacoid and up to $1\frac{1}{2}$ inches (4 centimeters) in length.

COPPER DEPOSITS.

The copper deposits at Red Butte are closely associated with the aplitic dikes. A few only of the more important prospects could be examined. At the Copper Queen and Metallic groups of claims, about 1½ miles southeast of the camp, a short tunnel and a few shallow pits have been opened on some bunches of oxidized ore containing cuprite, covellite, native copper, and chrysocolla, associated with hematite, limonite, and a little barite. The ore is rather irregularly distributed through a fine-grained aplitic dike, which in part is altered to a rusty jaspery material. Not enough work has been done to reveal the extent and trend of the ore. The general course of the dike with which it is associated appears to be nearly northwest and southeast. Whether the ore occurs throughout the dike or is confined to its sides, near the gabbro, was not clear from the exposures available at the time of visit. The openings also gave no clue to the original mineralogical character of the ore before its oxidation.

The Anaconda prospect, situated on the ridge of the same name, about a mile southwest of the camp and 400 to 500 feet higher, was being developed at the time of visit by open cuts. The workings are entirely within a broad north-south aplitic dike, which cuts fairly coarse gabbro. The valuable constituent here is native copper, which occurs along narrow seams and joints in the dike rock and is to some extent disseminated through it. Work had just begun and the extent of the ore body had not been ascertained. The rock in which the native copper occurs is not much altered and no quartz or other vein material is present.

At the Redeemer claim, about a mile north of the settlement, a fissure zone in the gabbro carries streaks of chalcocite partly altered to chrysocolla, malachite, and azurite.

Not enough prospecting has yet been done to determine the importance of the district as a source of copper. The superficial aspect of the deposits does not, however, suggest great size or richness.

Another belt of mineralization, not visited, is said to skirt the east base of the Jackson Range and to carry copper and antimony. At one place, a few miles southeast of Red Butte, some work was in progress at an antimony prospect in September, 1908, and it was reported that a shipment of partly oxidized stibnite would soon be made. Cinnabar also occurs a few miles southwest of Red Butte, west of the road to Rosebud. The deposit has been very little developed and was not examined by me.

HUMBOLDT RANGE.

GENERAL GEOLOGY.

The Humboldt Range (see Pl. I) has a total length of about 75 miles and attains a maximum altitude, in Star Peak, of nearly 10,000 feet. As all who have written about the range have recognized, it is divisible into two distinct parts, separated by a fault the position of which is marked by a low transverse pass known as Cole Canyon. The northern division, called by Louderback [a] the Star Peak Range, trends north and south and is about 32 miles long. The southern part, called by the same writer the Humboldt Lake Range, trends north-northeast and south-southwest and is over 40 miles long. The course of Cole Canyon is north-northwest.

According to Hague [b] the Humboldt Range consists of an Archean nucleus upon which rest Triassic strata of great aggregate thickness. These are overlain by Jurassic beds. There are also, on the lower slope, considerable masses of Tertiary rhyolite and basalt and a few rather small exposures of Miocene beds belonging to the Truckee formation.

When it is remembered that the geologists of the Fortieth Parallel Survey mapped an enormous area, that many of their lithologic determinations necessarily depended upon the color and erosional forms of rocks as seen from some commanding point of view, that their geologic field work was done before the topographic maps were available, and that the science of microscopical petrography was then in its infancy, no surprise need be felt that many and important changes must be made in their mapping by those who follow in their footsteps. Louderback [c] has shown that the so-called Archean nucleus, exposed in Rocky Canyon, southeast of Ryepatch, consists of a mass of post-Triassic intrusive granite with associated contact-metamorphic rocks. The intrusion, as Louderback observes, probably took place during the period of post-Jurassic deformation that affected the Great Basin region and the Sierra Nevada.

The Triassic rocks were divided by Hague into two formations. The upper or Star Peak formation is described as consisting of the following, numbered from the base up:

Tabular section of the Star Peak formation compiled from the description by Arnold Hague.

	Feet.
5. Quartzite and overlying limestone	4,000–5,000
4. Massive limestone	1,800–2,000
3. Black arenaceous slates	200– 300
2. Slaty quartzites alternating with greenish schistose rocks.	1,500
1. Limestones. Dark, almost black at the base, passing up into gray and blue varieties	1,200–1,500

[a] Basin range structure of the Humboldt region: Bull. Geol. Soc. America, vol. 15, 1904, p. 294.

[b] Descriptive geology, U. S. Geol. Expl. 40th Par., vol. 2, 1877, p. 714.

[c] Louderback, G. D., Basin structure of the Humboldt region: Bull. Geol. Soc. America, vol. 15, 1904, p. 318.

The total thickness is roughly estimated by Hague at 10,000 feet.

The Star Peak formation is noted for its abundant Middle Triassic vertebrate and invertebrate fossils, which have been described by Gabb,[a] Meek,[b] Hyatt and Smith,[c] and J. C. Merriam.[d] The stratigraphic section, however, is much in need of detailed study. Less hurried work will, I believe, make considerable changes in the sequence and lithology as interpreted by Hague and is likely to show some duplication of units in his section. Hyatt and Smith[e] state that the Upper Triassic is also represented in the Humboldt Range and list half a dozen fossils. The beds containing them are said to be unconformably overlain by limestone containing Jurassic forms.

The lower division of the Triassic was called by Hague[f] the Koipato formation, from the Indian name of the Humboldt Range. It is described as a series of quartzites and "porphyroids," with a total thickness roughly estimated at 6,000 feet. The "porphyroids," although recognized as closely resembling eruptive rocks, were regarded by Hague as "metamorphic products of the mixed quartz and feldspar rocks of the series of beds underlying the limestone of the Star Peak Triassic."

In the light of modern petrographic knowledge, Hague's descriptions and the accompanying chemical analyses of the "porphyroids" strongly suggest that they are not metamorphosed sediments, but are for the most part eruptive rocks. The exigencies of a hasty reconnaissance did not permit a thorough examination of the Koipato formation in 1908. Enough of it was seen, however, to leave no doubt of its dominantly igneous character. It consists of volcanic flows, mostly rhyolitic but including also andesitic lavas, associated with tuffs, conglomerates, grits, and limestones. The "porphyroids" are for the most part true igneous porphyries, although some of the tuffs may also have been included under this designation. As is to be expected in pre-Tertiary lavas that have been subjected to considerable deformation, the originally more or less glassy rocks are devitrified and altered, in some places even to being rendered rudely schistose. The extrusive and tuffaceous rocks are cut by rhyolitic and dioritic dikes that are also of early Mesozoic age, and the Koipato formation as a whole is a volcanic complex, of which nonvolcanic sediments, including limestones, form a subordinate part. No true quartzite was observed in the Koipato in the course of this reconnaissance.

a Paleontology, vol. 1, Geol. Survey California, 1864, pp. 19–35.

b Paleontology, U. S. Geol. Expl. 40th Par., vol. 4, pt. 1, 1877, pp. 99–129, Pls. X and XI.

c Triassic cephalopod genera of America: Prof. Paper U. S. Geol. Survey No. 40, 1905, pp. 21–23, Pls. XXII–XXV.

d Triassic Ichthyosauria: Mem. Univ. California, vol. 1, No. 1, Berkeley, Cal., 1908, pp. 18–19.

e Triassic cephalopod genera of America: Prof. Paper U. S. Geol. Survey No. 40, 1905, p. 26.

f Op. cit., p. 716.

Hague calls attention to the fact that the division of the Star Peak formation that I have numbered 2 on page 31 is very similar to the Koipato formation. An examination of these rocks near the Sheba mine, in Star Canyon, shows that they comprise andesitic and rhyolitic flows and tuffs, with some beds of tuffaceous grit and of limestone. All are altered and are slightly schistose parallel with the bedding planes. Were it not that these rocks appear to be underlain by a considerable thickness of black shaly limestone they would naturally be grouped with the Koipato. As it is, their stratigraphic position is a little doubtful. The plane of division between the Koipato and Star Peak formations is much in need of close study and accurate definition.

The structure of the Humboldt Lake Range has been carefully studied by Louderback,[a] who reached the conclusion that it is a block of folded pre-Cretaceous rocks elevated and tilted to the east in late Tertiary time by a zone of faults along the west front of the range. His explanation of the structure of the Star Peak Range is similar, and he believes that the main fault along the west base of this range curves to the east through Cole Canyon and has thus effected the separation of the Humboldt Range into its two distinct divisions.

The Humboldt Lake Range contains no important mines, and no examination of it was made during this reconnaissance. The older structure of the Star Peak Range was found to be broadly anticlinal as described by Hague, the axis of the fold trending north-northeastward from Cole Pass and thus crossing the range obliquely. From Cole Pass to Unionville is a belt of Koipato rocks 7 to 8 miles in breadth. Southeast of it is the Buffalo Peak mass of the Star Peak formation. Northwest of the Koipato belt the main ridge, from a point southwest of Unionville to the valley of the Humboldt near Mill City, is also made up of the Star Peak formation, overlain in the vicinity of Humboldt House by Jurassic strata. Thus the northern half of the Star Peak Range has the structure of a monocline dipping a little north of west. If faulting along the west base of the range, as deduced by Louderback, actually occurred, the consequent eastward tilting of the whole mountain block must have decreased by so much the dip of the older monocline, which is still the most conspicuous structural feature of this part of the range.

HUMBOLDT QUEEN MINE.

The Humboldt Queen mine is situated 5 miles northeast of Oreana, a now little-used station on the Southern Pacific Railroad, and is be-

---

[a] Op. cit.

tween the mouths of Sacramento and Limerick[a] canyons, at the west base of the Star Peak Range.

Little has been ascertained relating to the history of this mine. It is known to have been in operation in 1883, and the aspect of the empty buildings indicates that some work has been done during the past two or three years. No statement of the total production was obtained.

The general country rock of the mine is thin-bedded limestone, mapped by Hague as part of the Star Peak formation. It forms at this place a narrow north-south belt, bounded on the west by the sediments of Lake Lahontan and recent alluvium and on the east by granite porphyry or rhyolite porphyry, which on the Fortieth Parallel Survey map is included with the Koipato formation. The belt as exposed is from one-third to one-half mile wide and appears on the Fortieth Parallel Survey map as the narrow southernmost point of the area of the Star Peak formation that farther north makes up most of the range. The limestone beds are thrown into sharp complex folds trending generally about N. 10° E. A few of the folds are overturned, as illustrated in figure 2. Some beds are not noticeably meta-

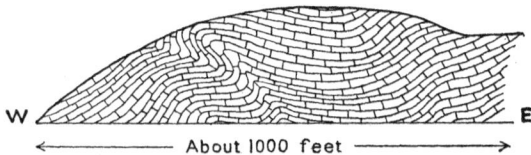

FIGURE 2.—Sketch of folded limestone as exposed on the slope of a small hill north of the Humboldt Queen mine.

morphosed, but others, especially the shaly ones, show considerable alteration, needles of tourmaline and rosettes of a dark-green mineral probably belonging to the chlorite group being noticed in some loose masses that had rolled down from the slope east of the mine. The limestone is cut by a north-south dioritic dike, in places several hundred feet wide, which passes a short distance east of the mine workings. The metamorphism may be due to the intrusion of this rock.

The Humboldt Queen mine comprises numerous openings along a section, 1,000 feet or more in length, of a complex north-south zone of veins in the limestone. The workings are mostly tunnels and shallow shafts or pits. At the south end of the explored zone, however, is an old inclined shaft from which most of the stoping appears to have been done, and just south of it a newer vertical shaft, apparently 200 feet or more in depth.

The veins are generally parallel with the bedding of the limestone and are sharply folded with that rock. In a few places they cut across the beds, but this appears to be exceptional. The consequence

---

[a] On the Fortieth Parallel Survey maps the name Sacramento is applied to the canyon up which passes the road from Humboldt River to Fitting or Spring Valley. This is now locally known as Limerick Canyon, the one north of it being called Sacramento Canyon.

of this conformity with the bedding is much variation in dip and strike, and of course the failure of some veins or parts of veins to extend to great depth. In some places the veins are horizontal, in others they are vertical, and the change in dip or strike is in many instances very abrupt, as may be seen from figure 3, in which some of the structural eccentricities of these veins are diagrammatically illustrated. At the collar of the old inclined shaft the vein stoped dips 55° E., but at a depth of 30 feet it curves and a short distance below it dips 60° to 70° W.

The topography of the vein zone is irregularly hilly, and as the intricate folding may bring the same vein to the surface at many places, it is impossible to say without detailed study how many bed

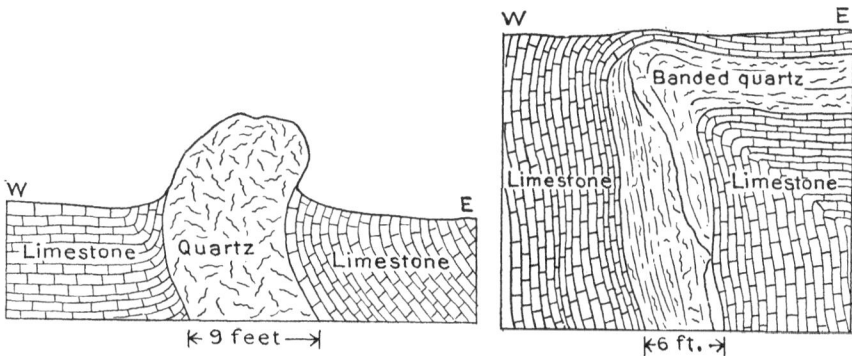

FIGURE 3.—Structural details at the Humboldt Queen mine.

veins there are in the belt. There appear, however, to be more than one.

The vein faces exposed show various widths up to about 10 feet. The filling is milk-white quartz, which in most places is banded parallel with the wall by thin dark seams of argillaceous material. The spaces now occupied by the veins were probably opened gradually, with successive deposition of quartz after each movement. Some of the folding appears to have taken place before all of the quartz was deposited, but much of the deformation is later than the veining, as the quartz is generally shattered, although not thoroughly crushed, and the banding of the quartz conforms as a rule to the folds. Other minerals noted in association with the quartz are calcite, pyrite, galena, sphalerite, and tetrahedrite.

FITTING AND AMERICAN CANYON.

The limestone belt in which is the Humboldt Queen mine terminates at Limerick Canyon. For about half a mile from its mouth this canyon is cut in the granite porphyry already referred to as lying east of the limestone. This rock was not closely examined, but has the appearance of an intrusive mass. East of it the road over Spring Valley Pass crosses for more than 6 miles the main belt of the Koipato formation, which here appears to be composed mainly of dark-weathering siliceous porphyries. The belt is probably wider than as shown on the Fortieth Parallel Survey map, which represents the spurs between American Canyon and Fitting as composed of the Star Peak formation, whereas they appear to be made up largely of Koipato volcanic rocks, angular conglomerates, and tuffaceous grits. The lower ends of these spurs are capped by basalt, which is not shown on the Fortieth Parallel Survey map.

Fitting, still better known by its old name of Spring Valley, and the settlement of American Canyon, about 2 miles to the south, were flourishing placer camps in the early eighties. The only lode mine of importance is the Bonanza King, situated about half a mile south of Fitting. This mine, formerly known as the Eagle, shipped ore to San Francisco at least as early as 1884 and milled the lower grades at Mill City. Afterward a 15-stamp amalgamating and concentrating mill was built at Fitting, but this was not successful. The mine was worked by a lessee in the winter of 1907–8, but was idle at the time of visit.

The workings comprise a vertical shaft 300 feet deep, with levels 300 to 400 feet long and of generally rectilinear plan. The 125-foot level was the only one examined, as the upper ones are stoped to the surface and the lower ones are under water.

The vein strikes generally N. 60° W. and dips 82° SW. It follows an altered dioritic dike, which is about 45 feet wide. This cuts the rocks of the Koipato formation, represented, in the part of the workings examined, by a porphyritic rhyolite. Other rocks noted in the vicinity of the mine, and probably cut at various depths by the dike and vein, are rather angular conglomerates, grits, and tuffaceous beds, all containing much rhyolitic material and some limestone.

The vein consists of firm banded quartz and is generally within the diorite, separated by a thin slab of that rock from the foot wall of the dike. The old stopes, now open to the surface, are 5 to 6 feet wide, with smooth regular walls. On the 125-foot level the stopes are in some places 8 feet wide. A notable feature of the vein is its displacement by cross faults, of which five or six are known in the workings. These strike nearly northeast, dip southeast, and have normal throws. The maximum offset effected by any of these faults is about 35 feet. They cut the vein sharply and contain some crushed vein matter and ore.

The ore is reported to have ranged up to $400 a ton, with a varying ratio of .gold to silver. It is not susceptible to treatment in the present mill, and the lessee stated that in 1907 he was able to recover by milling only $8 a ton on what the smelter returns showed to be $80 ore. The ordinary ore shows galena, pyrite, and sphalerite in milky translucent quartz, which in some places is slightly stained with copper carbonates. The richest ore contains in addition argentiferous tetrahedrite.

The placer deposits in Spring Valley are no longer worked. In American Canyon, however, about a dozen white people are living in the former Chinese village, and the gravels are being explored by two or three new shafts. For a distance of fully 2 miles from its mouth American Canyon shows signs of former activity in hundreds of pits amid piles of gravel. These pits are all that remain of the Chinese shafts, which were very small and ranged from 40 to 100 feet in depth. As little timber as possible was used, and after the productive gravel layers had been worked by drifting for short distances from a shaft it was abandoned and a new one was sunk. According to Mr. W. G. Adamson[a] the pay streaks were found at depths of about 40, 60, and 85 feet. The total alluvial deposit in the canyon is in places much thicker, the Midas shaft, put down some years ago, being reported to be 200 feet deep and all in gravels and clays. No gold, so far as known, has ever been taken from the bed-rock surface in the bottom of the canyon, although this appears to have been reached in one of the newer shafts estimated to be about 100 feet deep. The gold is said to have been found mainly at the bottom of gravel layers, underlain by seams of clay. The gravel shows little rounding and apparently is subangular and imperfectly assorted wash from the adjacent hills.

According to Mr. Adamson, particles of cinnabar were often found with the gold in the rockers, and some good-sized bowlders of this mineral are said to have been taken from the gravels. This led to prospecting for cinnabar lodes, and some development work has been done at two places on the north side of the canyon. The only one of these visited is 2 or 3 miles from the mouth of the canyon, upstream from the placer ground. The deposit is a soft, crushed, kaolinized zone in porphyritic rhyolite. No walls are exposed and the zone is at least 6 feet wide. It dips north into the hill at 15° to 20°. Little specks of cinnabar are scattered through the kaolin, but no ore has been found, although an incline 200 feet long has been sunk and considerable drifting done.

The second deposit, said to be more promising, is a short distance north of the settlement of American Canyon, in limestone, with some eruptive rock forming the hanging wall. The vein is said to strike northwest and to dip 20° SW. Specimens shown me by Mr.

---

[a] Oral communication.

Adamson contained abundant cinnabar in small irregular fissures in dark-gray limestone. The vein is said to have been traced for a length of 3,500 feet and to have been opened to a maximum depth of 150 feet.

UNIONVILLE AND VICINITY.

Unionville is situated on the east slope of the Star Peak Range, in Buena Vista Canyon. It is about 10 miles north of Fitting and 15 miles south of Mill City. Although no longer a bustling mining town, the quiet little settlement, with its running water, shade trees, orchards, and gardens, possesses charms that are the more pleasing because unlooked for in so generally arid a region.

The principal mines are a mile or two southwest of Unionville and at least 1,000 feet higher up the range. The rocks in the vicinity of the town and exposed along the road up to the mines belong to the Koipato formation and are conglomerates, grits, and limestones, with much siliceous porphyry, most of it probably rhyolite. The igneous rocks appear to occur both as flows and intrusions, but no careful examination of them could be made in the time available. The topmost member of the Koipato seen on the road to the mines is a sheet of porphyritic rhyolite apparently a few hundred feet thick. This rock is considerably silicified and extremely hard. In most places it shows flow banding, which, while more or less contorted, conforms generally to the dip of the mass as a whole and to the beds above and below it. In places, also, the rock is spherulitic and the microscope shows that it is a partly altered, originally glassy rhyolite flow. Much rock of this character appears to have been mistaken for quartzite by the Fortieth Parallel Survey geologists from the fact that its outcrops, seen from a distance, have some resemblance to that material. Overlying this rhyolite flow are thin-bedded, fossiliferous Middle Triassic (Star Peak) limestones, and it is in these that the ore bodies occur. Although generally gray in weathered exposures, the limestones are nearly black in underground workings, and are in part shaly. They form an elongated spoon-shaped synclinal mass, a little more than a mile in length and about 300 feet in greatest thickness, that occupies the summit of a hill on the spur between Buena Vista and Cottonwood canyons. The general relations of the limestone and rhyolite are roughly shown in figure 4, which is a mere sketch with no claim to accuracy of detail or to conformity to scale. Along part of its east side the limestone is bounded by a fault plane, along which it has been dropped against the rhyolite. A smaller mass of limestone, southeast of the larger one, is similarly faulted down along its west side. This contains the Wheeler mine.

The Arizona deposit is a bed or blanket vein that conforms with the bedding of the limestone and lies approximately 25 feet above the base of this rock or the top of the rhyolite. The vein has a maximum thickness of about 6 feet and averages about 3 feet. It is remarkably regular and persistent and appears to be practically continuous under the entire hill. The largest angle of dip observed is about 30°, and under the crest of the hill the deposit is nearly horizontal.

The old Arizona workings are at the north end of the hill, overlooking Buena Vista Canyon, and were started on a prominent outcrop of white quartz. The old stopes extend south from the surface into the hill for over 1,000 feet and have an area of about 18 acres. Practically no timber was used, and the stopes, still in excellent condition, constitute a great labyrinth of galleries between the neatly piled masses of waste that support the roof. Raymond,[a] who saw these stopes in 1871, noted the unusual basin-like form of the deposit and remarked that the conditions were more suggestive of a coal mine than of one worked for precious metals. It is a fact of some interest that deposits of this character were known and worked in Nevada at the time when our mining laws of 1866 and 1872, apparently so little applicable to them, were passed. The statement made by Shamel,[b] that "it is probable that at

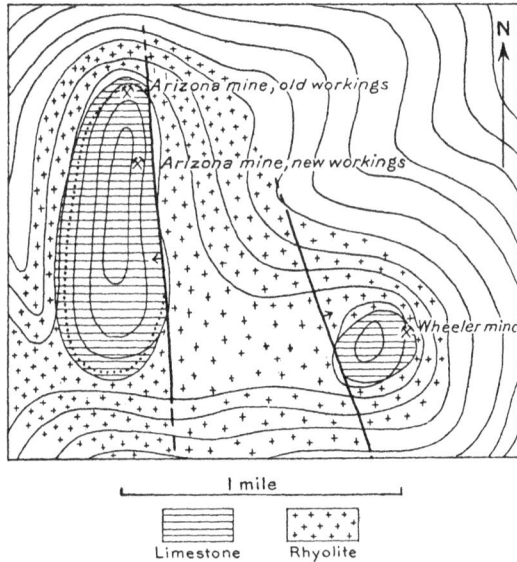

FIGURE 4.—Sketch plan showing the geologic relations of the Arizona mine.

the date of the passage of the statute of 1872 no instance was known of a vein which assumed a horizontal or even an approximately horizontal direction," is obviously incorrect.

The newer workings, which are developmental, are south of the old stopes. There are two tunnels, which connect with extensive exploratory drifts, crosscuts, and winzes or raises.

Although the vein as a whole has the form of a spoon it exhibits some structural complications. Faults, striking N. 20° to 30° W. and dipping west, step the vein down to the west. The throw is generally greater at the north, the maximum displacement observed being about 75 feet. At one place near the south end of the workings there is a slight reverse throw on one fault plane that farther north is associated with normal dislocation. The principal faulting is along the

a Statistics of mines and mining, Washington, 1872, p. 206.
b Mining, mineral, and geological law, New York, 1907, p. 233.

eastern edge of the syncline, and here the vein has been dropped against the rhyolite. In the vicinity of this fault zone both the vein and its inclosing limestones are, as a rule, disturbed. The vein is crushed, flexed, and to some extent dragged along the fault planes, some of this dragged material being good ore. The limestones are crumpled and in some places are mashed and squeezed to a structureless mass. In general, the contact of the vein with its walls is close and without gouge. In the vicinity of the faults, however, vein and limestone have slipped past each other, and in a few parts of the mine the vein disappears and its place is taken by a seam of gouge.

The normal vein material is solid milk-white to light-gray quartz. It generally shows more or less banding parallel to its walls and in some places is divided by thin partings of limestone. There is no evidence of any important replacement, and the quartz has crystallized in open spaces produced by the separation of the limestone along one or more bedding planes. The banded structure suggests that the separation was effected by successive movements, each followed by the deposition of quartz. There is a little calcite with the quartz, but it is rare. The recognizable ore constituents are pyrite, galena, sphalerite, and tetrahedrite. As a whole the quartz is not particularly rich in sulphides, and the tendency of these is toward fine dissemination along layers parallel with the walls of the vein. Some of the gray argentiferous particles, especially where the ore has been fractured and secondarily enriched, appear to be argentite; and some of the richest ore, yielding on assay up to 1,750 ounces of silver and 0.5 ounce of gold, contains dull, greenish-black masses that are apparently mixtures consisting largely of argentite.

The richest shoots of ore have probably been mined out, but according to Mr. Carmichael[a] the average of a large number of samples from all exposed faces of the vein is about 20 ounces of silver and 0.05 ounce of gold.

At the south end of the hill, and probably on the same vein as the Arizona, is the Nevada-Union mine, which produced about $60,000. It has been long abandoned and was not visited. Southeast of the Nevada-Union is the Wheeler (formerly the Henning) mine, on a vein which, although in a block of limestone now separated by faulting and erosion from the beds containing the Arizona vein, was probably once continuous with that deposit. The Wheeler mine has always been under the same ownership as the Arizona. It was worked by flat stopes from the surface and produced probably from $100,000 to $150,000.

In Cottonwood Canyon, 3 miles south of Unionville, is the Manoa or Pfluger mine, which has been worked in a small way for many years. The principal adit is 800 feet long and connects with several hundred

---

a Oral statement.

feet of drifts. The general country rock is rhyolite, which appears to be overlain by a rhyolitic conglomerate with a steep dip to the west. Lower down the canyon and stratigraphically below the rhyolite are volcanic (largely rhyolitic) conglomerates, grits, and shaly limestones. All these rocks were included in the Koipato formation by the Fortieth Parallel Survey geologists. The rhyolite is cut by a complex north-south basalt dike, and both rhyolite and basalt are much faulted and sheared along north-south lines. In some places this zone of disturbance is fully 400 feet wide. The general dip of the zone appears to be about 70° W., but this is not clearly shown and it is not known whether this is the same or greater than the dip of the associated beds.

The ore occurs irregularly as bunches and streaks in this sheared and faulted zone, especially near the basalt, and is for the most part a replacement of the crushed rhyolite by galena, sphalerite, tetrahedrite, and possibly some silver sulphantimonite, in a gangue of barite and quartz. The mine has never been an important producer.

In Jackson Canyon, between Cottonwood and Buena Vista canyons, is an antimony deposit in rhyolite that is one of the flows in the Koipato. The vein strikes N. 35° W. and dips 75° SW. It has a maximum width of about a foot, but is irregular. It is accompanied by some parallel sheeting and veining of the rhyolite. The vein consists of rather vuggy white quartz with unevenly distributed bunches of stibnite. Three tunnels have been run on the vein and some narrow stopes opened. No work was in progress in 1908 and it is not known whether any shipments have been made.

STAR CANYON AND VICINITY.

The site of the former town of Star City, whose history has already been briefly told, is 6 miles north of Unionville and about 12 miles southwest of Mill City. When visited in 1908 no mining whatever was in progress in the canyon, and although the greater part of the Sheba workings was accessible, very little could be seen of the De Soto mine, which is south of the Sheba on the opposite side of the ravine.

The rocks in the immediate vicinity of the mines are rather thin-bedded gray limestones and tuffaceous sandstones interbedded with thin flows of rhyolite and related rocks—the porphyroids of the Fortieth Parallel Survey reports. These volcanic rocks are considerably altered and are in places partly schistose, so that the original character of some of them is not altogether clear. Most of them are rhyolite or rhyolite flow breccia. Some, while containing abundant phenocrysts of alkali feldspar, have no primary quartz and may be trachytes. No glass remains in any of these rocks, and quartz, calcite, sericite, and chlorite are common as alteration products. This heterogeneous assemblage of sediments and flows is overlain, just above the mines, by the massive gray limestone of Star Peak, and is apparently underlain, below the Sheba mill, by dark slaty limestones in which the stream has excavated a more open valley than at the mines. All the rocks mentioned were placed in the Star Peak formation of the Triassic by the geologists of the Fortieth Parallel Survey, although, as noted on page 31, the abundance of volcanic material at the Sheba mine suggests the possibility of their correlation with the Koipato formation. The general strike of the beds and flows is N. 15° E. They dip 50° W.

The Sheba bonanza was close to the surface and, as the stopes and tunnels were carried into the hill, proved to be large and irregular. It consisted in part of several distinct lenticular seams of ore that followed certain bedding planes in the limestone or between the limestone and the associated grits and volcanic rocks. In many places these seams were connected by a network of veinlets across the limestone beds, and where these cross veinlets were abundant the whole mass was stoped as ore. The limestone was the rock most fractured and was the ore bearer. As a rule, fissures crossing a bed of limestone end at the plane separating this bed from grit or rhyolite.

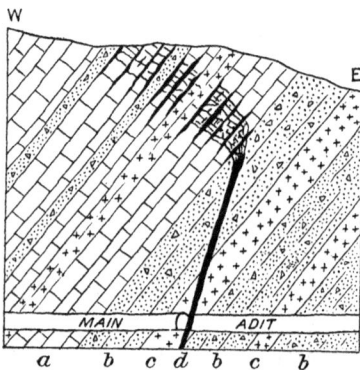

FIGURE 5.—Diagrammatic section of the Sheba ore body. a, Limestone; b, tuff; c, porphyry; d, ore.

The original pay shoots, which ramified through an area over 300 feet long from north to south and about 200 feet wide, were soon exhausted, and after vain attempts to find any deep continuation of the ore the Sheba mine was for a time abandoned. About the year 1871, however, it was ascertained that the pay shoots first stoped were connected with a well-defined and regular fissure vein. It was hoped that a new era of prosperity was at hand, but after the vein had been worked to an additional depth of rather more than 100 feet below the old stopes operations were discontinued. So far as is known, there has been no exploration of the vein below the main tunnel. The general structural relations of the deposit are diagrammatically shown in figure 5.

The main tunnel has its portal at the bottom of the canyon and runs N. 55° W. for about 800 feet through tuffs and flows to the vein, which on this level is in rhyolite. The vein strikes nearly north and dips from 65° to 75° W. It is up to 3 feet wide, consists of solid quartz with bunches of sulphides, and has been stoped on this level for about 160 feet, but apparently was here of low grade. Cutting obliquely across the layers of tuff and volcanic rock, the vein, as shown in figure 5, maintains its regularity up to the level above the main tunnel, a distance of about 100 feet. A short way above this level it enters limestone and at once changes in character. The stopes widen to 20 to 30 feet, the vein splits, and the whole deposit becomes more irregular and, as indicated in figure 5, has a general dip across the bedding to the east.

The Sheba mine produced an antimonial silver ore consisting of white quartz carrying argentiferous jamesonite, galena, sphalerite, pyrite, and tetrahedrite. Possibly other minerals rich in silver, such as stephanite and argentite, were present in the best ore, but these were not observed in the material now visible. Jamesonite is particularly abundant. According to B. S. Burton,[a] who analyzed the minerals many years ago, the jamesonite from the Sheba mine contains over 6 per cent of silver and the tetrahedrite from the De Soto mine 14.5 per cent of silver.

During the last operation of the mine the ore was carried by an aerial tramway about half a mile down the canyon to the mill, which is equipped with rolls, Huntingtons, jigs, and concentrating tables.

The De Soto mine, which was worked in 1861 and at various times since, is immediately south of the Sheba mine and is on the same zone of mineralization. There are two tunnels. The upper one, which is very devious, extends about 600 feet into the hill, measured in a straight line. The lower one, not safely accessible at present, is about 900 feet long. The rocks are in general the same as in the Sheba mine, and the ore occurs similarly along bedding planes and in fractured limestone. Most of the lenses of ore along the bedding planes appear to be connected with one or more veins that cut the beds. The bed veins have been stoped for distances up to 100 feet from these transverse fissures, which apparently are branches of the now filled channels through which the ore-bearing solutions rose. The ore of the De Soto mine is similar in mineralogical character to that of the Sheba.

About a mile down the canyon from the Sheba mine, on the south side, is a quartz vein that cuts black shaly limestone and carries stibnite. A tunnel has been run in on the vein, but apparently the deposit was never productive.

In Bloody Canyon, a mile or two south of the Sheba mine, is another stibnite deposit, which has been worked in desultory fashion for many years and from which some shipments of good antimony ore have been made. It was not visited.

RYEPATCH MINE.

On the west slope of the Star Peak Range, 5 miles west of Unionville and 4 miles east of Ryepatch station, is the old Ryepatch mine, which after producing ore of the currently reported total value of

---

a Contributions to mineralogy: Am. Jour. Sci., 2d ser., vol. 45, 1868, pp. 36–38.

over $1,000,000 has lain idle for more than twenty years. Originally known as the Alpha or Butte mine, the property was sold about 1872 for $80,000, and took its present name at the time of that transaction.[a]

The workings, comprising several tunnels and extensive stopes, are on the north side of Panther Canyon, in thin-bedded, more or less altered limestones supposed to belong to the Star Peak formation. The metamorphism, probably due to the granitic intrusion of Rocky Canyon referred to on page 31, is not here very conspicuous, consisting chiefly of the development of one or more species of fibrous silicates, probably belonging to the amphibole group, and in some beds of a white mica in almost microscopic scales.

The structural features of the deposit, which are unusual, may be most easily understood by reference to figure 6.

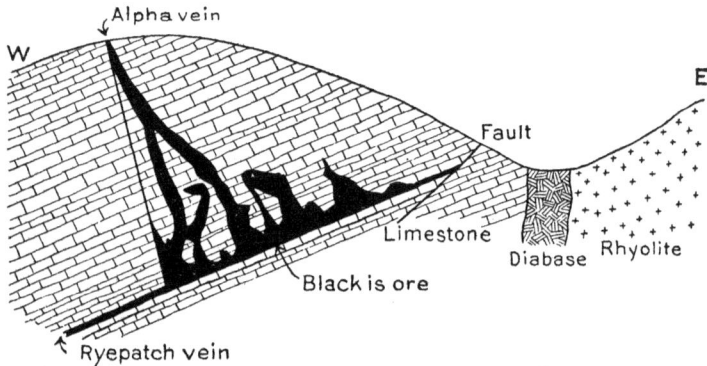

FIGURE 6.—Diagrammatic section of the Ryepatch ore body.

The Ryepatch vein, which follows the bedding of the limestone, has no recognized outcrop. Originally a shaft was sunk on a fault fissure that showed obscurely along the west side of a little lateral ravine eroded along a decomposed basic dike. (See fig. 6.) At a depth of 30 to 40 feet ore was found in a bed vein that strikes about N. 15° W. and dips 25° W. On the east this vein is cut off by the fault; on the west, at a distance of about 250 feet from the original discovery, it is joined by a fissure that strikes N. 25° E., dips 75° SE., and is known as the Alpha vein. All of the ore stoped came from the block of limestone between the two fissures. This mass is fissured in all directions and much of it is shattered to fragments. The ore, consisting of shattered limestone full of bunches and branching stringers of quartz and calcite, occurred as great irregular masses bounded in part by definite fissures but grading on most sides into country rock. The total length of the ore-bearing ground, from north to south, was about 600 feet and the total width about 250 feet.

---

a Raymond, R. W., Statistics of mines and mining, etc., for 1872, Washington, 1873, p. 156.

The Alpha vein itself is merely a fissure that bounds the ore on the west. It has a smooth slickensided hanging wall with practically no gouge and is not itself continuously filled with quartz. The Ryepatch vein, on the other hand, carried quartz and ore almost continuously from the point where it was first opened down to the junction with the Alpha fissure. In places this ore was narrow and had a definite hanging wall; elsewhere the ore extended from the foot wall for long distances and in very irregular fashion into the broken limestone, as roughly indicated in figure 6. The Alpha fissure, so far as known, does not extend below the Ryepatch vein. The Ryepatch vein continues westward beyond the Alpha fissure, but becomes smaller, more regular, and of much lower grade beyond the junction. A lower tunnel, some distance down the canyon from the main workings, has been run with the hope of cutting the Ryepatch vein. The vein, however, has not been identified at this depth. The tunnel follows a strong nearly north-south fissure zone, which suggests the possibility of considerable displacement of the Ryepatch vein by this or similar fissures west of the Alpha vein.

The vein material of the Ryepatch mine consists of quartz, calcite, pyrite, galena, sphalerite, tetrahedrite, and perhaps stephanite and argentite. The best ore is said to contain little or no calcite, galena, or sphalerite. No first-class ore was visible in the stopes at the time of visit.

ELDORADO CANYON.

The chief interest in Eldorado Canyon at present is derived from a deposit of cinnabar that was being prospected in 1908. The Ruby Cinnabar mine is about 6 miles southeast of Humboldt House (8 miles by road), on the northwest slope of Star Peak.

The ore occurs in a dark limestone, presumably belonging to the Star Peak Triassic, and has been opened by superficial tunneling. The beds in the vicinity dip generally west, but at the mine they are much disturbed and little was learned of their detailed structure at that place. The cinnabar is irregularly distributed through the fractured limestone, and the deposit as at present developed is a rolling, nearly horizontal mass about 80 feet long, 40 feet wide, and 6 feet thick. According to Mr. W. G. Adamson, the average tenor of the whole body is about 2 per cent of quicksilver. The cinnabar occurs as little branching veinlets, small bunches, and minute specks throughout the mass. In part it has filled fractures, but in part it has replaced crushed limestone. Some of the material through which the veinlets of cinnabar ramify is nearly black, earthy in appearance, and of noticeably high density. A lens shows that this material is crowded with minute specks of cinnabar and of pyrite or marcasite. The weight and dark color suggest also the presence of metacinnabarite, but verification of this supposition is not practicable in so fine grained a mixture of sulphides. The limestone adjacent to the ore is minutely fissured and faulted and contains many open crevices, especially under the ore body. Evidently there has been considerable removal of calcareous material by solution, and much of the fracturing associated with the deposit is probably due to the collapse of solution cavities and a settling down of the overlying rock. The workings have not exposed any vein or fissure that can be accepted as a probable main channel through which the ore constituents reached their present position.

NORTH END OF THE STAR PEAK RANGE.

North of Star Peak and Eldorado Canyon there has in the past been considerable mining activity in Humboldt, Prince Royal, Garden, Santa Clara, and other canyons. None of these old workings was visited. The Imlay (formerly the Morrison) mine, about 6 miles east of Humboldt House, is reported to be 300 feet deep with two levels. Work was in progress here in 1908 and the company was fitting up a mill near the mine.

On the crest of the range, at the head of Antelope Canyon, is the Thornton prospect, apparently in pre-Tertiary rhyolite. This was being developed in 1908, and a little rich ore was found containing free gold, associated with pyrite, sphalerite, galena, and a mineral resembling tetrahedrite, in a gangue of white quartz. Prospecting was going on also, in 1908, in Black Canyon, where the finding of some rich bunches of free gold ore in similar veins in the Triassic porphyries was causing a small influx of miners and speculators.

GENERAL FEATURES OF THE ORE DEPOSITS OF THE HUMBOLDT RANGE.

The ore deposits of the Humboldt Range, especially of the Star Peak Range, have certain common characteristics that mark their provincial unity or coherence. They occur in Triassic rocks, especially in the limestones of the Middle Triassic. Their age is not definitely known, but inasmuch as they appear to have been closely connected in origin with the folding of the Triassic and Jurassic beds it is probable that, like the gold veins of the Mother Lode belt in the Sierra Nevada, they date from early Cretaceous time. Mineralogically and structurally, however, they differ very widely from the California veins.

The ores, as a rule, are antimonial and contain much more silver than gold, with relatively little lead and zinc. Stibnite (or the similar lead sulphantimonite, jamesonite) is rather abundant, both with the silver ores and in deposits not containing notable quantities of the precious metals.

The tendency of the ores to deposit as bonanzas in certain structurally favorable places is noteworthy. It is exemplified by the bed veins of the Humboldt Queen, Arizona, and Wheeler mines and by the remarkable ore bodies of the Sheba, De Soto, and Ryepatch mines. In this connection also may be noted the close physical relation between the character of the ore body and the kind of wall rock and the fact that none of the deposits in the limestone has maintained its size and tenor when followed down into the underlying rocks. The absence of persistent veins at all comparable in richness with the masses worked within a short distance of the surface is one of the disappointing features of the region. The bonanzas apparently represent continued accumulation or enrichment, in favorable spots, of material that came in through channels insignificant in size or unsuited for the precipitation of rich ore.

PAHUTE RANGE.

GENERAL FEATURES.

The range east and south of the Humboldt Range is designated on the Fortieth Parallel Survey maps and on Spurr's[a] map as the Pahute[b] Range. Although this name is not in common use by the inhabitants of the region, it is here retained as a convenient term for the entire range, since the various names locally employed apply only to parts of the whole. Thus the northern part, east of the Humboldt Range, is commonly spoken of as the East Range; farther south it is the Table Mountain Range; and still farther south it appears on the Land Office maps as the Silver Range or the Stillwater Mountains.

From Humboldt River on the north the Pahute Range extends south for 50 miles, past Granite Peak, to the fortieth parallel. Here it bends and sweeps southwest for another 50 miles across the course of the Humboldt Range, and then again turning south continues to the vicinity of Wonder and Fairview, where it merges with other mountain groups.

The rocks of the Pahute Range are generally similar to those of the Humboldt Range. The two divisions of the Triassic are well represented, and the geologists of the Fortieth Parallel Survey have mapped some Jurassic beds along the west flank of the north end of the uplift. Exposed at several places over considerable areas, notably near Granite Peak, are masses of granite rocks, which, although referred to the Archean in the Fortieth Parallel Survey reports, are

a Bull. U. S. Geol. Survey No. 208, 1903, Pl. I. b Or Pah Ute.

at least in part post-Triassic. It is doubtful whether any Archean rocks occur in the range. Large areas, especially southeast of Spaulding Pass, north of Sou Springs, and in the vicinity of Table Mountain, are covered by Tertiary volcanic flows and tuffs. In the course of the present reconnaissance the range was crossed at only two places, and it is consequently impossible to add anything of value to what has already been published on the general structure. The Mesozoic rocks are much folded and the axes of the folds are not related in any regular manner to the topographic axis. As Louderback [a] has shown, the form of at least a part of the range has a much closer dependence upon late Tertiary faulting than upon the post-Jurassic folding.

The mining districts are more sparse than in the Star Peak Range, and the country, especially south of Granite Peak, is less frequented and is roamed by herds of mustangs or wild horses. At the north end, about 10 miles northeast of Mill City and just south of Dun Glen Peak, is Chafey, formerly Dun Glen, in the Sierra district. Tiptop, a new camp that had not attained any importance in 1908, is just south of Chafey. South of Natchez Pass is Orofino Canyon, where some mining was done thirty to forty years ago, and Rock Hill Canyon, where are old worked-out placers. Gold Banks, not visited, is apparently situated close to the divide between Grass and Pleasant valleys, whether in the Pahute or Sonoma Range was not ascertained. It was reported that the camp, after a brief period of activity, had become very quiet in 1908.

In French Boys Canyon, north of Granite Peak, a little prospecting is going on in rocks shown on the Fortieth Parallel Survey map as Koipato Triassic. They are conspicuously metamorphosed to slates and schists, contain abundant epidote, and show very plainly the influence of the granitic intrusion to the south.

Kennedy, at the east base of the range, east of Granite Peak, is practically deserted and no work of consequence was in progress in 1908.

About 30 miles a little west of south from Kennedy, in Cottonwood Canyon near Table Mountain, are nickel and cobalt veins, and a few miles south of them some copper deposits. No work was being done at any of these in 1908.

Still farther south, on the west side of the range, is Coppereid, in the White Cloud district, where a long crosscut tunnel is being driven by the Nevada United Mining Company. The exact position of this camp, owing to the general topographic and geologic inaccuracy of the Fortieth Parallel Survey map in its vicinity, remains in doubt. It is in T. 23 N., R. 34 E., and is apparently near the place indicated

---

a Op. cit., pp. 322-327.

in Plate I. About 2½ miles north of it is Copper Kettle, where, it is reported, are some promising copper prospects, and 6 to 7 miles farther north, at the end of a low spur projecting into the valley west of Chataya Peak, is a deposit of magnetite, said to occur in greenstone, although the Fortieth Parallel Survey map, probably in error, shows Tertiary basalt at this place. According to Mr. John T. Reid, about 1,000 tons of this ore was shipped to San Francisco in 1893–4. A few miles south of Coppereid is a new prospecting camp called Shady Run.

CHAFEY AND THE SIERRA DISTRICT.

The new town of Chafey, which in the late summer of 1908 was rapidly covering the old site of Dun Glen, is 10 miles northeast of Mill City and about 20 miles southwest of Winnemucca. After the closing of the Auld Lang Syne mine many years ago a little desultory work continued near Dun Glen, especially on Munroe Hill, where the veins carry free gold near the surface. According to the excellent article by Wisker in the Mining and Scientific Press, referred to on page 14, what was known as the Hendra group of claims was bonded to H. W. Kent in 1905 for $10,000. Under his direction a vein that assayed well was cut in a short tunnel, but the discovery appears not to have been followed up. The ground was next bonded to Charles Harlowe for $30,000. He interested E. S. Chafey in the prospect, and the latter undertook to raise the necessary money to develop it. Mr. Chafey began work about June, 1908, and by September of the same year is said to have shipped enough ore to purchase the mine. When the camp was visited in September, 1908, it presented a scene of brisk activity. Buildings were going up on all sides, and freight teams and stages, almost buried in dust, were plying back and forth between the new town and Mill City.

The ravine to which the name Dun Glen appears to have been originally given runs northeast and southwest, and 2 miles south of the summit of Dun Glen Peak heads in a pass through which goes the road to Winnemucca. The length of the glen is about 4 miles, and the claims about which the present activity centers are all on the southeast side. At the northeast end of the line, just south of the pass, is the Auburn mine. About a mile southwest of it is the Auld Lang Syne mine. Both had been long idle at the time of visit. A mile southwest of the Auld Lang Syne is Chafey's mine, on the Mayflower claim, locally called the Black Hole. About half a mile southwest of this is the Golden Bell tunnel. South of the tunnel, across a small east-west ravine that opens just below the town of Chafey, is a spur known as Munroe Hill. Here are many small veins, upon which lessees were busily at work in 1908.

The rocks in the vicinity of the mines mentioned were mapped as Star Peak Triassic by the Fortieth Parallel Survey geologists, and the few hours spent in the district were not sufficient for critical revision of their stratigraphy. West of the town are dark slaty rocks supposed to be Jurassic. Similar slates underlie the town of Chafey and occupy most of the bottom of the glen, so that it is not clear from a brief examination of the district why the line separating the Jurassic and Triassic on the Fortieth Parallel Survey map should have been drawn west of Dun Glen so as to divide rocks so alike in general appearance.

As the visitor goes east from Chafey and climbs the slope to tne mines, he crosses the strike of the beds and passes first over dark clay slates that strike N. 15° E. and dip 80° W. Lenses of limestone appear in the slates as the mines are approached, and along the mineralized zone itself the slate and limestone are succeeded by a belt of altered igneous rocks that are for the most part flows and flow breccias more or less interleaved with the sedimentary rocks. The width of this predominantly igneous belt was not ascertained, but was reported to be about half a mile.

The prevalent kinds of igneous rock are rather dark gray and show small dull phenocrysts of plagioclase in a fine-grained groundmass. Many are mottled and suggest squeezed flow breccias; others appear to be glassy (vitrophyric) lavas that have been rendered partially schistose. The microscope shows that the plagioclase phenocrysts are the only original minerals now recognizable. The groundmass is made up of an aggregate of quartz, calcite, sericite, kaolin, and other secondary minerals, of which the proportions vary in different flows or facies. The rocks represent vitrophyric andesite flows and flow breccias that have been folded and compressed with the slates and have developed incipient schistosity. Cutting all the rocks mentioned are dikes of normal olivine diabase (dolerite) with typical ophitic texture. In the diabase the plagioclase and the augite, which is brownish red in transmitted light, are fresh, but the olivine is more or less serpentinized. These dikes show no evidence of compression subsequent to their solidification, and were probably injected after the folding.

The Black Hole workings, which are about 1 mile east of Chafey and at over 1,000 feet greater elevation, consisted early in September, 1908, of a tunnel about 200 feet long, partly on the vein, with stopes extending about 25 feet up to the surface. Exploration was in progress also through a new inclined shaft, 60 feet deep, with short drifts.

The vein, which is of solid banded quartz up to 6 feet wide, strikes N. 50° E. and dips 45° to 50° SE. It accompanies a diabase dike and cuts obliquely across the volcanic flows and associated slates. As almost no crosscutting has been done little could be learned of the character of the wall rock. At one place the foot wall is limestone.

The vein minerals are quartz. galena, pyrite, sphalerite, and native gold, some of the gold being partly embedded in the galena. The gold seen is pale and probably contains a good deal of silver; according to Mr. Chafey the ratio of silver to gold in the ore is generally as 2 to 1, by weight. The richest ore occurred in the oxidized part of the vein, but this is very superficial. A short distance northeast of the Black Hole tunnel the vein passes into one of the andesite flows and splits up into a zone of stringers. At one place an open cut shows such a zone over 16 feet wide, with slate on the foot wall. The richest ore from the Black Hole is hauled to Mill City and shipped. The lower grades are treated (1908) in an old 3-stamp mill at Chafey.

The Golden Bell tunnel, about 600 feet long, follows a regular vein, with an average width of approximately 4 feet. It resembles the vein at the Black Hole workings and is presumably the same, although no accompanying dike was noted. The foot-wall rock was not exposed at the time of visit, but the hanging wall is altered andesitic flow breccia. Some stopes have been opened above the tunnel and evidently some ore was shipped or milled. No work, however, was in progress in 1908.

The veins of Munroe Hill strike nearly north and south and are almost vertical. They are in partly schistose andesite and andesitic flow breccia, and one of them, at least, accompanies a diabase dike. The entire zone of veins is at least 300 feet wide and appears to represent a southern extension and splitting up of the vein worked at the Black Hole. At the Bishop lease on the May Muller claim some high-grade ore was being taken in 1908 from a vein on the east side of the zone. The vein is from 1 to 2 feet wide, strikes north, and dips 80° E. It has a diabase dike along the hanging-wall side. The ore, which was being taken from an open cut near the top of the hill, was oxidized and contained free gold, with probably halogen compounds of silver. A blue-green material occurring as specks in the ore and supposed by the miners to be bromide of silver proves to be chrysocolla. At another lease, distinct from the former, although known as the Bishop & Co. lease, a tunnel was being driven on the north side of the hill to cut some ore discovered above, and at many other places on the Munroe Hill lessees were sinking shafts or beginning tunnels. The ore thus far found on the hill occurs near the surface and is probably considerably richer in gold than that below the limit of oxidation.

The Auld Lang Syne mine was worked through three tunnels down to water level, and the quantity of material on the dumps and at the site of the old mill about a mile above Chafey indicates that large and productive stopes were opened. Where the stopes come to the surface an excellent section of the vein is exposed, which is illustrated in figure 7.

The vein zone is about 100 feet wide and contains a number of nearly parallel quartz veins and stringers, separated by silicified andesite. The strike of the lode is N. 5° W. and it dips, as a whole, 65° E. On the hanging-wall side is a regular diabase dike 25 feet wide. The principal vein stoped is, as shown in figure 7, near the hanging wall of the zone. This dike, notwithstanding its proximity to the veins, is not generally altered and contains considerable olivine that has escaped serpentinization. The principal mineralogical change in the andesite, as elsewhere in this district, is the development of secondary quartz and sericite, the resulting rock outcropping and weathering much like a hard siliceous rhyolite.

The mine being quite deserted, I was able to procure no information concerning the character of the Auld Lang Syne ore. Some of the material last thrown out on the dump shows much arsenopyrite, with some pyrite arranged in depositional bands in quartz.

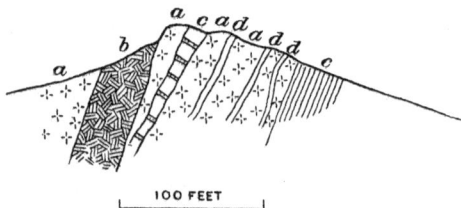

FIGURE 7.—Sketch section of the Auld Lang Syne vein zone. a, Silicified andesite; b, diabase; c, vein, stoped to surface; d, vein; e, slates.

Some of the veins near Chafey appear to be unusually persistent and regular. That they become lean within a moderate distance of the surface is suggested by all that can be learned of the history of the district; but one looking at the croppings of the Auld Lang Syne vein and at the work accomplished there finds it difficult to believe that the old mine will not some day be reopened, especially as its nearness to the railroad gives it and other mines in Dun Glen a great advantage over many in Humboldt County—as, for example, those at Kennedy, next to be described.

KENNEDY.

The almost deserted town of Kennedy lies at the east base of Granite (or Cinnabar) Mountain, about 30 miles by road southeast of Unionville and about 45 miles from the railroad at Mill City.

The distribution of the rocks in this vicinity is rather different from that represented on the Fortieth Parallel Survey map. The "granite" of Granite Mountain, supposed by the early geologic explorers to be Archean, is, at least locally, a diorite and is intrusive into Triassic rocks. Its outline, therefore, is undoubtedly much more irregular than they supposed. The bottom of the canyon from the crest of the range down to Kennedy is all in the diorite, and the rock appears to extend farther north in the vicinity of Kennedy than is shown on the Fortieth Parallel Survey map. On the other hand, the hills directly southwest of Kennedy, shown as "granite," are for some miles in that direction made up of altered pre-Tertiary (probably Triassic) volcanic rocks, with perhaps some sedimentary beds.

The diorite in the vicinity of Kennedy is, as usual in such intrusive bodies, not altogether uniform, and it is possible that much of the Granite Mountain stock may be granodiorite or some quartz-bearing granular rock near that type.[a] The rock in the pass west of Kennedy appears to be fairly representative of so much of the mass as was seen at close range. This is a bright gray, medium-grained rock showing abundant hornblende and biotite in a feldspathic base whose individual crystals are not distinct to the naked eye. As seen in hand specimens, the diorite is not noticeably quartzose.

The microscope shows that the feldspars are mainly oligoclase and mottled intergrowths of oligoclase or albite with orthoclase, the whole forming a rather intricately interlocking mosaic. Green hornblende is abundant, and the larger individuals as a rule inclose and are more or less intergrown with augite that is pale green in thin section. Both these minerals are intergrown with biotite and with magnetite, apatite, and titanite. A noteworthy feature of the diorite is the generally alkalic nature of the feldspars in association with rather abundant augite, hornblende, and apatite in a rock free from quartz. It is probably near monzonite in composition, and a chemical analysis would doubtless show less lime and more alkalies than in typical diorite.

The principal mine near Kennedy is the Gold Note, credited with a production of $60,000 to $70,000, but now idle, like all others in the district, where the only active industry in 1908 was the catching of wild horses for shipment to Oregon. It is situated on the south side of the canyon, about a mile west of town, and is opened by two crosscut tunnels. The lower and main adit runs S. 33° W. and is about 700 feet long. It cuts the vein, which strikes N. 65° W. and dips 25° (or less) S., about 375 feet from the portal. The general country rock is rhyolite, which is intruded rather irregularly by sheets and dikes of basalt. The rhyolite apparently forms thin flows, and the vein and part of the basalt have followed stratigraphic planes in the series of rhyolitic lavas. Associated with the rhyolite in some parts of the workings is a dark greenish-gray rock,

---

[a] Hague (U. S. Geol. Expl. 40th Par., vol. 2, 1877, p. 691) indeed describes the crystalline granitic rocks of Granite Mountain as made up chiefly of quartz and orthoclase, with scarcely any mica or hornblende (alaskite of Spurr). This suggests greater variation in the character of the rock than was evident from my own brief and limited observations.

evidently much altered, which was at first supposed to be metamorphosed limestone. The microscope shows, however, that it is an altered volcanic rock, probably andesite, and is composed chiefly of chlorite, calcite, and secondary quartz.

The basalt is a partly glassy, ordinary variety, in which some serpentinous material suggests the former presence of a small proportion of olivine. Its appearance under the microscope is that of an extrusive rock, whereas its relations underground indicate at least some intrusion. Clearly it was not intruded at any great depth.

The vein is from 1 to 3 feet wide and is composed generally of quartz and abundant pyrite. Associated with these minerals in the parts of the vein stoped are galena, sphalerite, tetrahedrite, and a little chalcopyrite. Along most of its course the vein is in rhyolite, but in some places it traverses basalt.

From the main adit drifts have been run in opposite directions. The west drift, apparently not very productive, follows the vein for about 90 feet to a zone of faulting that steps the gently inclined vein down below the level of the drift. These faults strike N. 20° W., dip west, and are normal. The first one drops the vein about 4 feet, and the second, 6 feet farther along the drift, carries the western continuation of the vein out of sight. Along the east drift the vein, at first in rhyolite, passes into basalt, becomes rather irregular, and at about 120 feet from the tunnel splits into two branches that diverge at a small angle. The north branch contains only bunches of ore. The south branch, which passes into rhyolite, has been stoped at intervals and apparently contained the principal ore bodies of the mine. At a distance of 500 feet from the adit the two branches of the vein are about 100 feet apart.

There was no one at hand in 1908 to give information about the mode of occurrence of the ore in the abandoned stopes, but it may be surmised that the pyritic parts of the vein are of low grade and that the richer ore is bunchy, difficult to mine on account of the low dip of the vein, and not amenable to treatment in ordinary mills.

The Borlasca mine lies southeast of the Gold Note and is apparently on the same vein or vein zone. The workings, which have not been productive, are all in the oxidized part of the vein and are shallow. The vein material is chiefly quartz and specular hematite, and as the unoxidized vein in the Gold Note mine is chiefly quartz and pyrite it is probable that the specularite was derived from the pyrite by weathering, although unfortunately the present workings afford no opportunity for observing the actual passage from pyrite to specularite.

In the bottom of the canyon, about half a mile above Kennedy, is the abandoned Hidden Treasure or K. and B. mine, in a rather basic facies of the diorite. An open cut exposes two veins about 18 inches wide separated by a foot or two of decomposed diorite. The veins strike northwest and dip about 50° SW. The vein material, as indicated by the dump of the closed main tunnel, contains very abundant pyrite and some sphalerite, with a subordinate quantity of quartz and calcite.

Half a mile north of Kennedy is the Imperial mine, which is credited with a production of about $10,000. It was not visited, but is said to be in the diorite and to have yielded lead-silver ore containing some gold. Some specimens of galena seen in Kennedy were stated to have come from this mine. Attempts were made to work this ore in a 20-stamp plate-amalgamation mill with cyanide tanks, now owned by the Borlasca Company. This and the various smaller mills in the district appear to have been designed without any regard whatever to the character of the ores to be treated.

NICKEL AND COBALT DEPOSITS OF COTTONWOOD CANYON.

From Kennedy to Boyer's ranch at the mouth of Cottonwood Canyon the distance by road is about 30 miles.[a] The usual route is through Pleasant Valley past the volcanic Sou Hills and Sou Springs, the latter being about halfway in the journey. These springs consist of six or seven circular pools, 40 to 50 feet in diameter, with funnel-shaped bottoms, distributed along the crest of a calcareous tufa mound that rises from 50 to 60 feet above the alluvium of the valley. The pools differ in temperature, some being cold and some at about 85° C. A description of these springs with a good illustration (looking south toward Boyer's ranch) and some chemical analyses are given by Hague.[b] From Sou Springs south the road skirts the east face of the Stillwater Range, which is here very precipitous and is notched by narrow ravines, many of which have not cut down to the main valley, but have built up high-angle alluvial cones at their mouths. The rocks in this part of the range appear to be mainly Triassic (Star Peak) limestones cut by masses of some light-colored intrusive rock.

The topography and geology in the vicinity of Cottonwood Canyon are very inadequately represented on the Fortieth Parallel Survey map, which shows the Triassic rocks ending at the canyon, with Tertiary rhyolite south of them, all being capped to the west by basalt.

The canyon for about half a mile from its mouth is cut through dark-gray and reddish indurated clay shales, overlain by gray limestone, and this in turn by several hundred feet of light-colored thin-bedded quartzite. All three rocks are intricately intruded and contorted by tongues, dikes, and irregular masses of dioritic rock, whose

---

a Boyer post-office is on most maps of Nevada erroneously placed at the south end of the great salt playa of Osabb (Dixie) Valley, whereas it is really north of it.

b U. S. Geol. Expl. 40th Par., vol. 2, 1877, pp. 704–705, Pl. XX.

complex relations to the sedimentary beds are impressively displayed in the steep bare front of the range northwest of Boyer's ranch. West of these beds, which are supposedly Triassic, the canyon for several miles is in a dioritic rock that shows considerable variation in texture and composition from place to place. Probably the most widespread variety is a medium to fine grained diorite consisting originally of plagioclase, hornblende, and perhaps augite, with considerable titanite and some apatite and magnetite. As a rule the diorite is much altered. The plagioclase is partly changed to calcite and sericite and the hornblende to epidote, calcite, chlorite, and a pale-green secondary amphibole. Other facies observed are darker and contain more hornblende, with some biotite. Still others are coarsely crystalline, are roughly foliated, and have a mottled appearance due to poikilitic development of the feldspar.

In the upper part of the canyon, in the vicinity of the nickel deposits, the diorite is intrusive into andesite and andesite breccia, much of which is so altered that its original character is scarcely recognizable. Near the diorite the andesite is silicified, carries disseminated particles and bunches of hematite with streaks of copper ore, and weathers in craggy rusty outcrops. Farther from the contact the rock is dark purplish gray and shows in places a suggestion of a tuffaceous or breccial structure and the outlines of feldspar phenocrysts. The microscope shows that the plagioclases, partly altered to sericite and calcite, are the only original minerals recognizable. The femic minerals have been changed to calcite, and particles of hematite are disseminated thickly through the groundmass.

Both the andesitic rock and the diorite are cut by dikes and by small irregular masses of a white feldspathic rock in which rutile in minute scattered crystals and grains is the only dark constituent. The grain of these dikes varies from that of an ordinary granite to that of a rather fine quartzite. The microscope shows that the plagioclase is chiefly sodic oligoclase, with a little orthoclase and probably some albite. The sharply angular spaces between the subhedral crystals of feldspar are occupied in part by calcite and in part by fine granular, microscopically radial aggregates of a mineral supposed to be quartz. The calcite is presumably secondary, as some of it has formed at the expense of the feldspars. Most of it, however, appears to have filled empty spaces by infiltration, the feldspars that inclose it preserving their original crystal outlines. The accessory minerals are rather abundant rutile with apatite and zircon. The rock is of uncommon character and will probably be briefly described in a separate publication when a chemical analysis of it is completed.

The work of the Fortieth Parallel Survey indicates that the sedimentary rocks and andesitic rocks described are probably Triassic and that the diorite and white dikes are late Mesozoic. No additional information was obtained in the course of the present reconnaissance.

Resting upon these rocks at the head of the canyon is a thick series of nearly horizontal Tertiary volcanic rocks. Near their base is at least 300 feet of andesitic breccia with some intercalated vesicular basalt. The actual base of this series was not observed. Overlying the andesite tuffs is about 200 feet of rhyolite, covered in turn by a succession of thin volcanic flows, which, as seen from a distance, apparently include both andesite and basalt. Table Mountain is capped by these flows.

Among the many interesting geologic features of Cottonwood Canyon to which but scant attention could be given in so hasty a visit may be especially mentioned an extinct hot-spring deposit 2 or 3 miles above Boyer's ranch and about half a mile below the Nickel mine. This is a great mound of siliceous sinter surmounted by a crater-like orifice fully 200 feet above the present stream, which has cut deeply into the mound and has exposed its steep quaquaversal stratification. The deposit rests on diorite and was formed at a time when the canyon had not attained its present depth.

The Nickel mine is situated on the north side of Cottonwood Creek, from 3 to 4 miles above Boyer's ranch. It was opened about 1882, when, it is reported, a car of 26 per cent nickel ore was shipped to Camden, N. J., and was worked for eight years. It was then closed in consequence of litigation. Work was resumed in 1904 but ceased again in 1907.

The workings are on a contact between diorite and andesite or andesite breccia, the rock being too much altered for certain identification. The contact, which is here more regular than elsewhere and may be due to local faulting, strikes N. 50° E. and dips 35° NW., the diorite forming the foot wall. The ore, which is all in the andesitic rock, has been exploited by cuts, tunnels, and an inclined shaft about 200 feet deep. It occurs in narrow fissures that make various angles with the contact and are not individually persistent. Some ore has been followed for a distance of 100 feet from the contact, measured perpendicularly to that plane. The diorite in the foot wall shows fissuring and disturbance and carries stringers of quartz, but no ore.

The seams of nickel ore are rarely over 3 or 4 inches in width, and all the material seen was more or less oxidized. The original filling, of which some residual masses remain, is partly a sulpharsenide of nickel, probably gersdorffite, according to Mr. W. T. Schaller, who tested the material in the Geological Survey laboratory. It is probable that chloanthite and other nickel minerals are also present, as the original ore has the appearance of being in part a mixture of arsenides or sulpharsenides of nickel. The residual kernels of sulpharsenide are veined and coated with a bright green hydrous nickel arsenate, probably annabergite, as determined by Mr. Schaller. This constitutes most of the ore. No quartz or other distinctive gangue mineral was noted in the veinlets. The nickel minerals are not confined to the major fissures, but have penetrated the rock in their vicinity for several inches along joints and microscopic cracks, forming a low-grade ore which the company has attempted to treat by leaching with sulphuric acid.

The Lovelock mine, about half a mile west of the Nickel mine, is reported to have shipped a total of about 500 tons of high-grade nickel-cobalt ore, but has long been idle. The workings comprise a labyrinth of superficial burrowings, by which the miners have followed or sought for the small erratic veinlets of ore, and a precarious shaft that no attempt was made to explore. This shaft is said to have been sunk to water and to be connected with exploratory drifts just above the water level. It apparently is not much more than 100 feet deep. The country rock is altered andesite like that at the Nickel mine. No diorite was seen, but it would probably be cut in deep workings.

The seams or veinlets of ore run in practically all directions and have no definite walls. The ore, all of which is partly or wholly oxidized, is more complex than that of the Nickel mine and contains copper and cobalt as well as nickel. The minerals recognized are tetrahedrite, erythrite (cobalt bloom), azurite, and green crusts that according to Mr. Schaller contain copper and nickel arsenates and sulphates and consequently may be a mixture of annabergite and brochantite.

A few miles south of Cottonwood Canyon, high on the east slope of the range, is a copper prospect which is now known as the Treasure Box, but which appears to have formerly been called the Bell Mare or Cornish mine. The ore, in the form of pyrite and chalcopyrite, is disseminated through andesitic tuff in the lower part of the Tertiary volcanic series. This impregnation extends through a belt at least a mile long and several hundred yards wide, which trends about N. 70° E. The most abundant sulphide is pyrite, the chalcopyrite occurring only here and there in bunches. The pyritization is by no means uniform and fades out indefinitely into the surrounding rock. Apparently there is no vein and no master fissure whence the mineralization has emanated.

At the east end of the deposit some oxidized copper ore has been taken from open cuts and a shaft has been sunk, apparently without the discovery of workable ore. A water-jacket furnace was erected at this place, the highest point at which the deposit outcrops, but was never used. Two tunnels, one of them about 400 feet long, have been run near the west end of the pyritic zone.

COPPEREID.

Coppereid, in the White Cloud district, on the west slope of the Stillwater Range, was the most southern locality visited in the course of the reconnaissance. The only work in progress in 1908 was by the Nevada United Mining Company, which is driving a tunnel into the south side of White Cloud Canyon at an elevation of about 1,000 feet above the valley of Carson Sink.

The lower part of the canyon is a narrow, steep gorge cut through a moderately coarse, slightly micaceous granite, intruded by many dikes of granite porphyry and by a few of diorite porphyry. Farther up the granite appears to grade into granite porphyry (although this was not established with certainty) and the porphyry is succeeded to the east by a series of limestones and calcareous shales, with a few beds of gypsum, into which it is plainly intrusive.

These sediments are overlain to the east by volcanic flows and tuffs that form the crest of the range, although on the Fortieth Parallel Survey map this is represented as being made up of the Star Peak Triassic. No fossils are known from the sedimentary rocks in White Cloud Canyon, but they are presumably Middle Triassic. They are metamorphosed by the granite porphyry intrusion, the purer limestones being marmorized and others being altered throughout by the development of garnet, epidote, fluorite, quartz, axinite, specularite, and metallic sulphides.

The workings of the Nevada United Mining Company comprise a main crosscut tunnel near the bottom of the canyon. The course of this is S. 37° E., and its length at the time of visit was 2,000 feet. On the steep hillside above the tunnel are old shallow workings that were operated about fifteen years ago for oxidized copper ore, which was reduced in a small smelter at the mouth of the canyon. Still higher, about 850 feet vertically above the main tunnel, are two tunnels known as the Twin tunnels. Both are at the same elevation, have their portals near together, and run nearly in the same general direction, so that they apparently represent an effort to explore the ground with as much labor as possible. One of these penetrates the hill for 700 feet.

About 350 feet above these tunnels is the summit of the ridge separating White Cloud Canyon from the next ravine to the south. Here is a large outcrop of specularite, which carries in places a little oxidized copper ore. There is probably from 100,000 to 200,000 tons of this material actually exposed on the hilltop and in the shallow exploratory workings that have been run into the mass. Much of it is practically pure specularite, ranging in texture from soft, greasy, rougelike varieties to coarsely foliated kinds. The copper is rather sparingly distributed through some parts of the iron ore as malachite. There are several of these masses of iron ore on the ridge. They appear to occur as pods, lenses, and irregular bodies associated with considerable fluorite and distributed through a broad mineralized zone that strikes east and west and dips about 45° N. The masses themselves strike and dip in various directions. On the west the zone, which is in limestone, ends at the intrusive contact of this rock with the granite porphyry a few hundred yards from the great hematite outcrop on the crest of the ridge.

The Twin tunnels were run under the supposition that the hematite on the ridge is the gossan of a great vein. They penetrated a little oxidized ore consisting of chrysocolla, specularite, limonite, calcite, colorless fluorite, and epidote, but disclosed nothing of value and did not cut any mass of specularite comparable with that above.

A similar expectation led to the driving of the main tunnel. The portal of this is in white to buff crystalline limestone carrying small bunches of specularite and in places much garnet and some pyrite. The beds appear to strike southeast and dip northeast, but the structure is very indistinct. At 400 feet from the portal a crosscut was driven east for 600 feet; and from this, about 200 feet in, a branch was run north to a broad zone of marmorized limestone heavily impregnated with slightly cupriferous pyrite. In places this is so abundant as to form nearly solid masses. This is probably the same zone that carries specularite near the mouth of the main tunnel and the one from which some of the oxidized copper ore was mined near the surface fifteen years ago. There is no recognizable vein; the pyrite has formed in the limestone by metasomatic replacement and is without definite boundaries. Beyond the branch the east crosscut continues for 400 feet through much broken and disturbed limestone, with many open crevices partly filled with soft, earthy limonite. The material suggests active solution by oxidizing waters, the formation of cavities, and the subsequent collapse of the cavernous rock into a mass of coarse, angular rubble.

The greater part of the main tunnel is in fine-grained dark limestone and originally calcareous shales that have been converted into dense siliceous hornstones. These are cut by many small veinlets carrying calcite, axinite, sphalerite, pyrite, and chalcopyrite. In some places they contain round or ellipsoidal concretions up to 6 inches in diameter, which when broken open show a septarian structure and a kernel of silicates, quartz, calcite, pyrite, pyrrhotite, chalcopyrite, and sphalerite. Among the silicates epidote, in short prisms and grains thickly embedded in calcite, is most abundant. It is associated with brushes of capillary crystals that have not been certainly identified but may be an amphibole—possibly tremolite.

According to a letter from Mr. John T. Reid, manager of the mine, the tunnel on January 18, 1909, was 2,750 feet in length and was making so much water from the face (1,250,000 gallons in twenty-four hours) that work had been temporarily suspended. Prior to tapping this water the tunnel passed through a number of small fissures carrying some chalcopyrite, and about 100 feet from the face, according to Mr. Reid, went through 3 feet of vein material containing chalcopyrite and pyrrhotite.

The conclusion reached from the short examination made of the main tunnel and the old workings, supplemented by a general survey of the surface, is that all the workings are in a zone of pronounced contact metamorphism, within which bunches of lean sulphides and of specular iron of irregular shape are rather erratically distributed. These show a decided tendency to form thick pods or lenses of indefinite outline, rather than distinct and persistent veins of which the positions could be calculated for depths far below their outcrops. Thus deep crosscutting involves more than the usual hazard of such a mode of exploration. While one would be rash to assert that no profitable vein will be cut in the tunnel, there is nothing on the surface or in the old workings that demonstrates the existence of any but isolated masses of ore throughout a contact zone.

The small quantity of primary ore thus far found, consisting of pyrite, pyrrhotite, and chalcopyrite, is of low grade, as is to be expected in deposits of contact-metamorphic type. The old stopes near the surface, from which ore was formerly obtained for the smelter at the mouth of the canyon, show considerable migration and concentration of oxidized copper ore, and solutions that extracted their copper from lean sulphides (or possibly from sulphides previously enriched) have deposited chrysocolla by direct replacement of crystalline limestone. The steps of this process, from narrow veinlets of chrysocolla widened by replacement, through rounded kernels of limestone inclosed in the copper silicate, to solid masses of the latter, are beautifully shown in the old stopes.

The occurrence of great masses of specularite on the top of the ridge and of large bodies of pyrite far below in the main tunnel suggested at first that the specularite was derived from pyrite by oxidation. Specularite is present, however, although not so far as known in large masses, in the lower tunnel, where it has crystallized with sulphides and garnet as a primary contact mineral. No material could be found that showed the passage of sulphides into specularite, and the evidence at this locality, while not conclusive, rather favors the view that all of the specularite is of direct contact-metamorphic origin.[a]

---

[a] Since this was written considerable masses of specularite intimately associated with pyrite have been cut in the tunnel 2,800 feet from the portal and about 1,000 feet below the surface. This places the contact-metamorphic origin of the specularite beyond reasonable doubt.

SONOMA RANGE.

The Gold Run district, in which the most important property is the Adelaide mine, 11 miles nearly due south of Golconda, is on the east slope of the northern part of the Sonoma Range.[a] The rocks of this part of the range are generally similar to those of the Pahute and Humboldt ranges. No attempt was made in this reconnaissance to study their lithology or structure, except at the Adelaide mine, where they have been mapped as Star Peak Triassic by the geologists of the Fortieth Parallel Survey.

The district was organized in 1866. Development apparently was slow, for in 1870 the principal shaft, the Golconda, was only 80 feet deep. South of this were the Cumberland, 50 feet deep, and the Jefferson, with still shallower workings. There were some small mills in the district, and desultory attempts were made to work the partly oxidized ores up to about 1897, when the Glasgow and Western Exploration Company acquired the mines and 15 claims along the ore-bearing zone. This company built 12 miles of narrow-gage railway from Golconda to the mine and erected a smelter and concentrating mill at the junction of its road with the Southern Pacific Railroad. This plant, consisting of two roasting furnaces and three reverberatory smelting furnaces, with the ordinary arrangement of crushing and concentrating machinery, was operated for a time on ores from Battle Mountain and from Adelaide, and some matte was shipped. The process, however, proved unsuited to the Adelaide ore and was abandoned. A few years ago the mill was remodeled and 120 concentrating tubes of the Macquisten type were installed. An interesting description of this remarkable plant has been given by W. R. Ingalls,[b] and from this the reader may obtain some idea of the ingenuity, simplicity, and effectiveness of this novel process, in which the heavy sulphides are floated off while the gangue minerals sink. Some improvements in the first installation were in contemplation in 1908, and the mill was in use by Mr. Macquisten solely for experimental purpose. Its total capacity was given as 120 tons in twenty-four hours. It produced when in full operation a 20 per cent concentrate from 2.7 per cent copper ore, leaving about 0.2 per cent in the tailings. The weakest point in the process appears to be in the relatively low recovery from the slimes.

The main shaft of the Adelaide mine, 300 feet deep, is situated on the south side of Gold Run Creek, close to the site of the old settlement of Cumberland. The general country rock is dark calcareous slate, within which is a layer or series of beds of limestone from 50 to 75 feet in total thickness. This bed strikes north and dips 65° E.

---

a The Havallah Range of the Fortieth Parallel Survey reports.
b Concentration upside down: Eng. and Min. Jour., vol. 84, 1907, pp. 765-770.

This limestone layer carries the ore, which in some places occupies the full width from one slate wall to the other, although as a rule the zone contains horses of altered limestone that is nearly free from sulphides. The ore body is undoubtedly large and has been extensively stoped above the 100-foot level for 400 feet without any indication of a diminution in size. Below this level, which is approximately at the bottom of the zone of partial oxidation, exploratory drifts have been run at vertical intervals of about 50 feet, revealing abundant ore. The bottom level was under water at the time of visit.

The ore is a metasomatic replacement of the limestone and consists of pyrrhotite, chalcopyrite, sphalerite, and a little galena, in a gangue of garnet, vesuvianite, diopside, calcite, orthoclase, and a very little quartz. Common pyrite is probably not altogether absent, although it does not appear in the specimens of ore collected. The presence of orthoclase is uncommon in this mineralogic association, but adularia has been noted by Spurr and Garrey [a] in the altered limestones of the Velardeña contact zone. At Adelaide the orthoclase is poikilitic and contains inclusions of vesuvianite, garnet, diopside, and quartz. The ore is definitely bounded only where it is in contact with the slates. Elsewhere it merges gradually and irregularly into limestone containing silicates but very little of the sulphide constituents. A banding of the limestone, due to alternations of silicate and calcite layers, is common, particularly near the ore, and the bands in places are contorted and crumpled. As a whole the ore is of low grade, averaging about 3 per cent of copper; but the quantity available appears to be large, and the difficulties in the way of its successful concentration and treatment will probably soon be overcome.

The present workings do not afford much evidence of secondary enrichment. The old stopes between the 100-foot level and the surface were in mixed sulphide and oxidized ore, but whether chalcocite was present in quantity is not known.

About 600 feet north of the main shaft, on the opposite side of tne little creek, is a tunnel that runs north in the ore zone for 2,000 feet. For a distance of 500 to 600 feet from the portal the tunnel is in ore. Beyond this the limestone zone is generally lean or barren, although there are a few bunches of ore near the face and some stopes above the tunnel were formerly worked from a now abandoned shaft on the hilltop.

A notable feature of the Adelaide ore bodies, in view of the fact that the nearest area of eruptive rock (mapped as granite on the Fortieth Parallel Survey map) is fully a mile east of the mine, is their close correspondence to ores of typical contact-metamorphic deposits. The granitic rock was not examined in 1908. For at least a quarter of a mile east of the mine the rocks are dark clay slates alternating

---

[a] Ore deposits of the Velardeña district, Mexico: Econ. Geology, vol. 3, 1908, p. 708.

with thin-bedded limestones. All are much crumpled but maintain a generally east dip and are on the whole much less metamorphosed than the limestone beds in which the ore occurs. It is probable that an intrusive mass underlies the sedimentary rocks at the Adelaide mine, and that the hot mineralizing solutions rose along what is now the ore zone, in consequence of favorable fissuring in this particular belt of limestone.

West of the mine the slopes, seen from a distance, show many outcrops suggestive of rhyolitic porphyries, which accord in general with the mapping of the higher part of the range by the geologists of the Fortieth Parallel Survey as Koipato Triassic. Within these a number of prospectors are developing veins that carry some gold and silver. None of these prospects was visited.

MINERALOGY OF THE ORE DEPOSITS.

Introductory statement.—For convenience of reference the minerals noted in the ores or closely associated with them are here given in alphabetic order, with brief notes on their occurrence. The list is obviously not an exhaustive one for the region, which contains many deposits not visited.

Amphibole.—A fibrous mineral, not certainly identified, but closely resembling tremolite, occurs with epidote, vesuvianite, garnet, and sulphides in the metamorphosed calcareous rocks at Coppereid. A similar fibrous mineral was noted in some of the altered limestone at the Ryepatch mine.

Annabergite.—A bright-green, hydrous nickel arsenate, probably annabergite, is an important constituent of the ore of the Nickel mine in Cottonwood Canyon, west of Boyer post-office.

Argentite.—Sulphide of silver has been found in shallow workings in rhyolite at Rosebud, associated with kaolinite, limonite, and jarosite. Presumably it was present also in some of the rich silver ores mined in former days near the surface in the Humboldt Range. A specimen of ore seen in Unionville and said to have come from the Arizona mine apparently contains argentite.

Arsenopyrite.—The sulpharsenide of iron was noted only in material on the dump of the Auld Lang Syne mine, near Chafey, associated with pyrite and quartz.

Axinite.—Axinite, a complex borosilicate of calcium, aluminum, and other bases, occurs in the altered calcareous shales of the Coppereid contact zone.

Azurite.—The blue hydrous copper carbonate is nowhere abundant in the region examined, but is present in small quantity in the Red Butte copper district and at the Lovelock cobalt-nickel mine, in Cottonwood Canyon.

Barite.—Sulphate of barium was noted as a gangue mineral in ore from a prospect near Fitting; with galena, sphalerite, tetrahedrite, and quartz in the Manoa or Pfluger mine, near Unionville; and in some of the copper prospects at Red Butte.

Bornite.—Minute specks of a mineral resembling bornite were noted with chalcopyrite in quartz at the Mazuma Hills mine, in the Seven Troughs district. Not enough of the material was obtained, however, for a satisfactory determination.

Brochantite.—Brochantite, a green basic sulphate of copper resembling malachite, probably occurs in the cobalt-nickel ore of the Lovelock mine.

Calcite.—Calcite is sparingly present with quartz in the ores of the Humboldt Queen and Arizona mines, in the Humboldt Range, in some of the veins near Kennedy, and in the deposits of contact-metamorphic type at Coppereid and Adelaide.

Cerargyrite.—Cerargyrite, or horn silver, is generally an inconspicuous mineral and probably occurred in many of the rich silver deposits formerly worked in the Humboldt Range.

Chalcocite.—Copper glance, cuprous sulphide, was noted only at Red Butte, in a vein in gabbro. It is probably present also in the upper parts of the ore bodies at Adelaide.

Chalcopyrite.—Copper pyrite is an important constituent of the Adelaide ore, associated with garnet, vesuvianite, epidote, diopside, calcite, quartz, orthoclase, pyrrhotite, sphalerite, galena, and pyrite. It is present also in the cobalt-nickel ore of the Lovelock mine; in the disseminated pyritic deposits of the Treasure Box mine, southwest of Boyer's ranch; in some of the veins near Kennedy; and sparingly in the Mazuma Hills mine, at Seven Troughs. Close search would probably show it to be present in small quantities also in most of the mines of the Humboldt Range.

Chloanthite.—A mineral of the smaltite-chloanthite group is perhaps present with gersdorffite at the Nickel mine, in Cottonwood Canyon, but it has not been definitely identified.

Chrysocolla.—The hydrous silicate of copper occurs in the oxidized zone of the Copperoid contact deposits, where to some extent it has metasomatically replaced limestone. It is sparingly present at Red Butte and is found in minute specks in some of the oxidized ore of Munroe Hill in the Chafey district.

Cinnabar.—The red sulphide of mercury occurs in fractured limestone in Eldorado Canyon, in the Humboldt Range; in a vein in limestone and in kaolinized rhyolite in American Canyon, in the Pahute Range; and at an unvisited prospect on the edge of the Black Rock Desert, a few miles south of Red Butte. It has also been found in the gold-bearing gravels of American Canyon.

Copper.—Native copper was noted only near Red Butte, where it is disseminated through aplitic dikes that cut gabbro.

Covellite.—The deep blue cupric sulphide occurs sparingly as a product of weathering in the superficially opened copper prospects of Red Butte.

Cuprite.—Earthy cuprite occurs with covellite, chrysocolla, and iron oxides at Red Butte.

Diopside.—Monoclinic calcium-magnesium pyroxene occurs with garnet, vesuvianite, orthoclase, and sulphides in the ore of the Adelaide mine.

Epidote.—Epidote, a basic orthosilicate of calcium, iron, and aluminum, is fairly abundant in the contact zone at Coppereid, associated with garnet, specularite, axinite, and sulphides.

Erythrite.—Cobalt bloom, a hydrous cobalt arsenate, is an important constituent of the ore of the Lovelock mine in Cottonwood Canyon. It is associated with tetrahedrite, azurite, and arsenates and sulphates of nickel and copper (probably annabergite and brochantite).

Fluorite.—Fluorspar is fairly abundant in the contact zone at Coppereid and is especially associated with specularite.

Galena.—Galena is widely distributed in the Humboldt region, and is found in the ores of nearly all of the districts visited. None was seen, however, in the Seven Troughs district. Specimens from the Imperial mine near Kennedy showed enough galena to constitute a lead ore, but as a rule the mineral is disseminated rather sparingly through quartz with sphalerite, pyrite, and tetrahedrite. At Chafey free gold and galena are directly associated. In Star Canyon galena and jamesonite occur together.

Garnet.—Brown garnet is abundant at both Coppereid and Adelaide, where it occurs with other silicates and with sulphides in altered limestone.

Gersdorffite.—The principal unoxidized constituent of the ore of the Nickel mine, in Cottonwood Canyon, is, according to Mr. W. T. Schaller, probably gersdorffite, a sulpharsenide of nickel. It is associated with a green alteration product, which he has determined as annabergite.

Gold.—Native gold occurs in visible masses, some of them of unusual size, in the quartz veins of the Seven Troughs district. It is also found associated with quartz and galena in the Chafey district and in various prospects in the Humboldt Range near Star Peak. Specimens of free gold have been found also in rhyolite just south of Golconda, but not in such form or quantity as to induce deep mining.

The principal placer localities of the region are American Canyon and Spring Valley (Fitting), on the east side of the Star Peak Range, and Rock Hill Canyon, in the East Range. Several other canyons, especially in the northern part of the Star Peak division of the Humboldt Range, also yielded some placer gold.

Hematite.—Large masses of specularite occur in the contact-metamorphic zone at Coppereid, and the same mineral is abundant in the superficial workings of the Borlasca mine, near Kennedy. In the latter place the specularite is possibly gossan material derived from the oxidation of pyrite. Some cryptocrystalline siliceous hematite occurs in the Red Butte district, associated with the copper ores. The material, which is in part jasperoid, appears to have been formed by alteration of parts of the aplitic dikes.

Jamesonite.—Jamesonite, a sulphantimonite of lead with considerable resemblance to stibnite, is an abundant constituent of the ore of the Sheba and De Soto mines, in Star Canyon. It is associated with galena, sphalerite, tetrahedrite, and pyrite in quartz. Both the jamesonite and the tetrahedrite, as shown by Burton,[a] are silver bearing.

Jarosite.—Jarosite, a hydrous sulphate of alkalies and iron analogous in its formula with alunite, was identified microscopically as minute yellow crystals associated with kaolinite and limonite at the Brown Palace mine, near Rosebud.

Magnetite.—Masses of iron ore, chiefly magnetite, occur about 20 miles southeast of Lovelock, at the northwest base of the Pahute Range. The deposits, said to be in "greenstone," were not visited.

Malachite.—The green hydrous carbonate of copper is present in small quantities at Red Butte, Coppereid, Adelaide, Cottonwood Canyon (Stillwater Range), and probably at other localities where ores containing copper minerals have undergone oxidation.

Orthoclase.—The potassium feldspar orthoclase occurs with garnet, vesuvianite, diopside, calcite, and sulphides in the ore of the Adelaide mine. The mineral is xenomorphic and poikilitic, containing inclusions of the minerals associated with it.

Proustite.—Ruby silver has been reported from the Seven Troughs district. None was seen in 1908.

Pyrrhotite.—Magnetic pyrite is an abundant constituent of the ore of the Adelaide mine and occurs also in the contact zone at Coppereid.

Silver.—Native silver is said to have been found in Wildhorse Canyon, in the Seven Troughs district.

Sphalerite.—Zinc blende is a very widely distributed mineral in the Humboldt region and occurs in nearly all of the districts visited. It was not observed, however, at Seven Troughs. It is not as a rule abundant at any one place, but, like galena, is disseminated rather sparingly through the various ores.

---

[a] Contributions to mineralogy: Am. Jour. Sci., 2d ser., vol. 45, 1868, pp. 36–38.

Stibnite.—The widespread distribution of stibnite, or antimony glance, is characteristic of the region covered by this reconnaissance. Deposits of stibnite have been worked in the Trinity or Antelope Range, southeast of Red Butte; in the Humboldt Range, southeast of Lovelock; in Star Canyon; in Bloody Canyon, 2 miles south of Star Canyon; in Jackson Canyon, about a mile south of Unionville; and in the Bernice district, about 16 miles south of Boyer's ranch, where there was at one time considerable activity. The Bernice district is south of the area studied in 1908. Stibnite occurs also associated with some of the gold ores at Seven Troughs, and is undoubtedly present at many localities not examined in the course of this preliminary investigation. The mineral occurs generally in fissure veins with quartz, although at Seven Troughs it forms bunches in soft, crushed rock or gouge.

Sulphur.—Native sulphur is extensively mined at the Rabbit Hole sulphur mine, about 5 miles northwest of Rosebud.

Tetrahedrite.—Gray copper, or tetrahedrite, like sphalerite and galena, occurs in most of the silver-gold deposits of the Humboldt and Pahute ranges. It is nowhere very abundant nor in large masses, but is disseminated in specks or small bunches through the predominant quartz gangue. As a rule it is argentiferous (freibergite) and its presence indicates an ore rich in silver. The tetrahedrite of the De Soto mine, according to Burton,[a] contains 14.5 per cent of silver.

The mineral was noted in the ores of the Humboldt Queen mine; of the Arizona, Wheeler, and Manoa mines, near Unionville; of the Sheba and De Soto mines, in Star Canyon; of the Ryepatch mine; and of the Gold Note mine, near Kennedy. It also occurs with cobalt minerals at the Lovelock mine.

Tourmaline.—Tourmaline was not observed in close association with ores, but occurs as an abundant microscopic constituent of altered limestone or calcareous shale near the Humboldt Queen mine.

Vesuvianite.—The characteristic contact-metamorphic mineral vesuvianite, a basic calcium-aluminum silicate of uncertain formula, occurs abundantly with garnet, diopside, calcite, quartz, orthoclase, and sulphides at the Adelaide mine. It was identified by its optical properties as seen under the microscope.

TYPES OF DEPOSITS REPRESENTED IN THE REGION.

Basis of classification.—The ore deposits of a given region may be classified in many different ways, as, for example, (1) by geologic age, (2) by form, (3) by supposed origin, (4) by mineralogic character, or (5) by essential metallic contents. In the present paper an attempt

---

[a] Loc. cit.

will be made to group the deposits primarily with reference to the closely related fourth and fifth bases of comparison. The grouping thus effected will to some extent coincide with classifications dependent upon genetic or morphologic features.

Antimonial silver deposits.—A large proportion of the deposits of southern Humboldt County consist essentially of silver ores carrying varying minor quantities of gold. These ores are prevailingly antimonial, the silver being combined chiefly in tetrahedrite or jamesonite. They generally contain in addition a little galena (probably argentiferous) and sphalerite, with of course some pyrite. The gangue is quartz, and as a rule the sulphides are subordinate to the gangue and are rather finely disseminated through it. Argentite and other rich silver-bearing minerals may occur in the upper parts of some of these deposits.

To this class belong the deposits of the Sheba and De Soto mines, in Star Canyon; of the Arizona, Wheeler, and other mines near Unionville; of the Humboldt Queen and Ryepatch mines; of the principal mines at Kennedy and Fitting; and of a number of unvisited mines and prospects in the Humboldt Range. Possibly the deposits of the Rosebud district belong in this group, but too little ore has yet been found there to furnish a safe basis for comparison.

With the exception of the doubtful ores at Rosebud, all of these deposits are in Triassic rocks and most of them are in the Star Peak formation. Beyond the fact that the ores are post-Triassic their age is unknown, but it is thought probable that they are pre-Tertiary and were deposited during or after the post-Jurassic intrusions and folding that affected the whole Great Basin region and the Sierra Nevada.

The antimonial silver ores generally fill fissures, but they show much variety in form. Some deposits, as that of the Bonanza King mine at Fitting, are comparatively simple, nearly vertical veins. Some, while generally of simple tabular form, have a low angle of dip, as in the Gold Note mine, at Kennedy, or are nearly horizontal, as at the Arizona mine, near Unionville. Others are of irregular shape and are related in special ways to fissuring and bedding, as in Star Canyon and at the Ryepatch mine. Still others, such as the Humboldt Queen deposits, are sharply folded bed veins.

None of these deposits has been worked to great depth, and the rich ores appear to form bonanzas within short distances from the surface. As a rule, the deeper workings show a decrease both in the size and tenor of the deposits. At present there is no production from any mine in ore of this type, although many of them were extensively and profitably worked in the sixties.

Gold-silver deposits.—The deposits that owe their value chiefly to gold are those at Seven Troughs and at Chafey. There are marked differences, however, between the ores of the two districts. Those at

Seven Troughs are in Tertiary volcanic rocks; those at Chafey are in Mesozoic volcanic and sedimentary rocks, probably Triassic. At Seven Troughs the gold is coarse, occurs in lodes made up of many small irregular quartz veinlets, and is accompanied by comparatively little pyrite, with as a rule no other sulphides. Stibnite, it is true, occurs in some of the mines, but not, according to my observation, as a constituent of actual ore. At Chafey the native gold is in smaller particles and occurs in well-defined solid veins in which the quartz carries considerable galena, less sphalerite, and comparatively little pyrite. Whether the veins of the Auld Lang Syne and other abandoned mines exhibit the same character is not known, although arsenopyrite appears to have been the principal sulphide in the lower levels of the Auld Lang Syne. Neither at Seven Troughs nor at Chafey were the accessible workings deep enough in 1908 for a thoroughly satisfactory comparison of the two kinds of deposits.

Gold prospects, of which the ores resemble the Chafey ore, in that the gold is associated with galena and sphalerite in quartz veins traversing Triassic rocks, have been opened at various places in the region, particularly in the vicinity of Star Peak.

Copper deposits.—Like the gold deposits, the copper deposits of the region fall into two classes. One of these is exemplified by the deposits southwest of Boyer's ranch, in Tertiary andesite, and by those at Red Butte, which are in igneous rocks doubtfully regarded as of Tertiary age. The deposits near Boyer's ranch are diffuse pyritic disseminations. Nothing is known of the Red Butte deposits beyond what may be seen from very shallow workings in oxidized material; the greater part of the copper appears to be disseminated through altered aplitic dikes in gabbro, and the native metal may be expected to give place in depth to some sulphide, perhaps to chalcopyrite.

The deposits at Coppereid and Adelaide, on the other hand, are in calcareous sedimentary rocks, probably belonging to the Triassic. They have the mineralogical characteristics of contact-metamorphic deposits, although at Coppereid only is there visible relation between the metamorphism and a mass of intrusive rock that effected the alteration. Garnet, chalcopyrite, pyrrhotite, sphalerite, and pyrite are common to both localities, although no sulphide ore bodies have yet been opened at Coppereid. Axinite, fluorite, epidote, and specularite occur in the contact zone at Coppereid, but were not noted at Adelaide. At the latter place, on the other hand, the altered limestone contains vesuvianite, diopside, and orthoclase. Doubtless a more thorough study would increase the lists of silicate minerals present at each locality. The deposit at Adelaide, like those at Yerington,[a] shows that ore bodies of essentially contact-metamorphic type

---

[a] Ransome, F. L., The Yerington copper district, Nevada: Bull. U. S. Geol. Survey No. 380, 1909, pp. 99-119.

are not necessarily in direct contact with igneous rocks near the surface.

Antimony and quicksilver deposits.—The general character and distribution of the deposits of stibnite and cinnabar have been briefly outlined in the preceding section on the mineralogy of the ore deposits. The occurrence of both kinds of ores at a number of widely separated localities is one of the noteworthy features of this part of Nevada, and indicates that the ore deposition, in spite of its varied local manifestations, has had some ultimate dependence upon conditions regional in their prevalence. In other words, the ores show a general provincial relationship.

The antimony and quicksilver deposits, with the exception of the stibnite at Seven Troughs, are all, so far as is known, in Triassic or Jurassic rocks and are supposedly of the same age as the antimonial silver-gold ores. No facts are known, however, that absolutely rule out a Tertiary age for some of these deposits.

Nickel and cobalt deposits.—The nickel and cobalt deposits in Cottonwood Canyon consist of sulpharsenides of nickel (gersdorffite in part), tetrahedrite, and some compound of cobalt with sulphur, arsenic, or antimony, with the various oxidation products of these minerals. The ores fill small fissures in much-altered andesite or andesite breccia cut by diorite and may be genetically related to the intrusion of the latter rock. The age of the diorite is probably late Mesozoic. The age of the nickel and cobalt ores is not definitely determinable. If their deposition followed closely the intrusion of the diorite they are probably pre-Tertiary. On the other hand, the occurrence of copper both with the nickel and cobalt ores and in Tertiary andesite at the Treasure Box mine, a few miles south of Cottonwood Canyon, suggests that the nickel and cobalt deposits may be Tertiary.

CONCLUSION.

The southern portion of Humboldt County is part of a metallogenetic province characterized chiefly by the prevalence of antimonial ores of silver with numerous and widely scattered deposits of stibnite and cinnabar. There are in addition some deposits of gold-silver, copper, and nickel-cobalt ores. Ore deposition probably began immediately after the intrusion of the Triassic and Jurassic sediments in late Mesozoic time by a magma of generally granodioritic composition, comparable with that which invaded the rocks of the Sierra Nevada at the same period, and continued into the Tertiary. The known Tertiary deposits are essentially gold-silver ores and copper ores, but it is possible that some of the other types are also Tertiary.

INDEX.

www.ingramcontent.com/pod-product-compliance
Lightning Source LLC
Chambersburg PA
CBHW070519200326

41519CB00013B/2852

* 9 7 8 1 6 1 4 7 4 0 3 9 1 *